1st Level Maths
Textbook 1A

T0173419

Series Editor: Craig Lowther

Authors: Antoinette Irwin, Carol Lyon,
Kirsten Mackay, Felicity Martin, Scott Morrow

© 2019 Leckie

001/24052019

10 9 8

ISBN 9780008313951

Published by
Leckie
An imprint of HarperCollinsPublishers
Westerhill Road, Bishopbriggs, Glasgow, G64 2QT
T: 0844 576 8126   F: 0844 576 8131
leckiescotland@harpercollins.co.uk   www.leckieascotland.co.uk

HarperCollins Publishers
Macken House, 39/40 Mayor Street Upper, Dublin 1, D01 C9W8, Ireland

Publisher: Fiona McGlade
Managing editor: Craig Balfour
Project editors: Alison James and Peter Dennis

Special thanks
Answer checking: Caleb O'Loan
Copy editor: Louise Robb
Cover design: Ink Tank
Layout and illustration: Jouve
Proofreader: Dylan Hamilton

A CIP Catalogue record for this book is available from the British Library.

Acknowledgements
Images © Shutterstock.com

Printed and bound by Grafica Veneta S.p.A.

Contents

Answers and free downloadable resources

Answers

All answers to the Before we start, Let's practise and Challenge questions in Textbooks 1A, 1B and 1C can be downloaded from our website here:

https://collins.co.uk/primarymathsforscotland

Free downloadable resources

There are free downloadable resources to support Textbooks 1A, 1B and 1C. These can be downloaded, printed out and photocopied for in-class use from our website here:

https://collins.co.uk/primarymathsforscotland

There are two types of resources:

- **General resources.** These are helpful documents that can be used alongside Textbooks 1A, 1B and 1C, and include, for example, blank ten frames, number lines, 100 squares and blank clock faces.

Ten frames

- **Specific resources.** These are supporting worksheets that relate to either a particular area of learning or a specific question and are labelled with a unique resource reference number. For example, 'Resource 1A_2.2_Let's practise_Q2' is a specific downloadable resource for Textbook 1A, Chapter 2.2, Let's practise Question 2.

Resource 1A_2.2_Let's practise_Q2

1.1 Estimating and comparing collections

We are learning to compare the size of collections by estimating.

Before we start

Can you draw this number of flowers?

12

We can sort collections in different ways to help us estimate which is bigger.

Let's learn

Look at these collections. Which collection has more – the teddies or the dinosaurs?

We can make it easier to estimate which collection is bigger by organising them like this:

The collection of dinosaurs is bigger.

1)

Put one counter onto every object.

Sort the fish and shell counters into two lines.

Which group is bigger, the fish or the shells?

2)

Use cubes and build them into towers to help you estimate.

Which group has more, red or green cubes?

3) Isla and Finlay are comparing how many sweets they have.

Can you estimate who has more? Use counting objects to organise them.

⭐ **CHALLENGE!** ······························

Draw between 10–20 circles on a piece of paper or a whiteboard for a partner. Do not tell your partner how many circles you have drawn.

Ten frames

Ask them to estimate how many circles you have drawn.

Now use counters and tens frames to work out how many circles there really were. Whose estimate was the closest?

1.2 Estimating and describing collections

We are learning to describe whether a number is closer to zero than 10 using materials.

Before we start

a) How many dots can you see?
b) How many more do you need to make 10?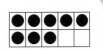

We can use ten frames or number lines to help us estimate whether a number is closer to 10 or zero.

Let's learn

Is the number 7 closer to 10 or 0?

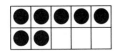

When there are more than five counters on the ten frame, the number is closer to 10 than 0.

We could use a number line too. Let's look at number 3.

We can see that 3 is closer to 0 than to 10.

1) Count out eight counters or cubes. Now put them onto a ten frame by filling up the top row first, then the bottom row.

Ten frames and Number lines

Is the number 8 closer to 10 or 0?

2) Count out four counters or cubes.

Now put them onto a ten frame.

Is the number 4 closer to 10 or 0?

3) Is the number 2 closer to 10 or 0?

Use the number line to help you estimate.

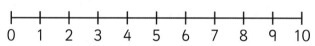
0 1 2 3 4 5 6 7 8 9 10

4) Is the number 9 closer to 10 or 0?

Use the number line to help you estimate.

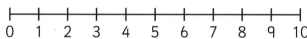
0 1 2 3 4 5 6 7 8 9 10

5) Use a ten frame or number line to help you estimate if 6 is nearer to 10 or 0.

CHALLENGE!

a) Nuria says that 14 is closer to 10 than 20.

Is she right?
Use ten frames or a number line to help you decide.

b) Can you find all the numbers between 10 and 20 that are closest to 10?

1.3 Estimating and grouping collections

We are learning to estimate amounts by grouping into fives and tens.

Before we start

Amman says the number that comes before 50 is 59. Nuria says the number is 49. Who is correct and why?

We can group collections into fives and tens to help us estimate how many there are.

Let's learn

Look at this collection of teddy bears:

We can sort them into rows of five using a ten frame:

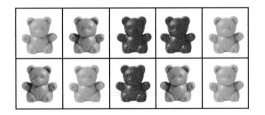

Now it is easier to estimate.

We can see that there are 11 teddy bears in this collection.

1) You will need blank ten frames for this question.

Estimate how many cars there are. Put a counter over each of these cars on the page. Now put all the counters onto your blank ten frames.

Ten frames

2) Look at these ten frames.

a) How many balls are there?

b) How many cars are there?

c) How many cats are there?

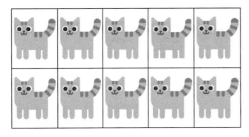 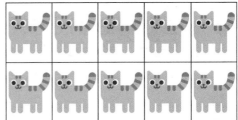

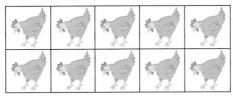

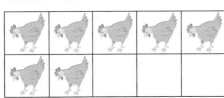

d) How many chickens are there?

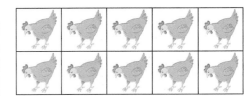

Estimation and rounding

3) Look at this picture.

Ten frames

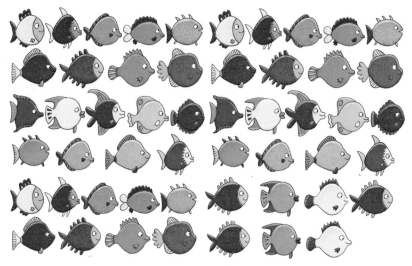

a) Estimate how many fish there are.

b) Put a counter over every fish. When you have done this, move each counter from a fish onto your tens frames.

c) Use the tens frames to work out how many fish there really are.

CHALLENGE!

Amman needs 100 pencils. He has this many so far.

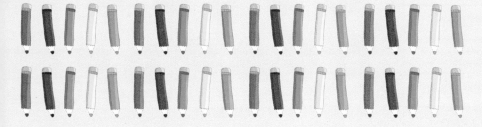

a) Estimate how many more pencils he will need to make 100 in total.

b) Count the pencils to check your estimate.
   Was your estimate close?

2.1 Reading and writing numbers (1)

> We are learning to read and write the numbers from 0 to 20 in words.

Before we start

Isla draws three ducks. She writes 3 next to her picture.

Draw and write:

a) five cats　　b) eight balls　　c) two apples　　d) seven snakes

Let's learn

> Numbers can be written in numerals and in words.

Talk about these number words.

11	eleven
12	twelve
13	thirteen
14	fourteen
15	fifteen
16	sixteen
17	seventeen
18	eighteen
19	nineteen
20	twenty

eleven

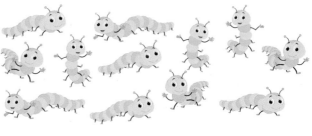

twelve

1) Write the numbers that come out of the number machines in your jotter.

a) 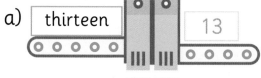 thirteen → 13

b) 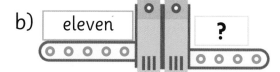 eleven → ?

c) 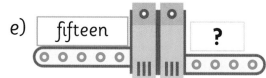 nineteen → ?

d) 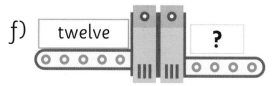 twenty → ?

e) 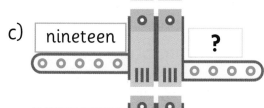 fifteen → ?

f) 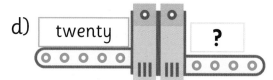 twelve → ?

2) What colour is the monster? One has been done for you.

a) seventeen – **pink**   b) twenty   c) eighteen
d) fourteen   e) twelve   f) eleven

| 17 | 18 | 14 | 20 | 11 | 12 |

3) How many baby monsters? Write the number word in your jotter.

CHALLENGE!

Fix Finlay's spelling.

a) 11 – elevin   b) 15 – fiveteen
c) 13 – threeteen   d) 12 – twelv

2 Number – order and place value

2.2 Counting in tens

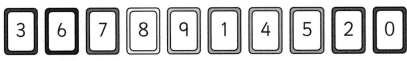

We are learning to count forwards and backwards in tens.

Before we start

a) Write these numbers in order in your jotter. Start with zero.

| 3 | 6 | 7 | 8 | 9 | 1 | 4 | 5 | 2 | 0 |

b) Count backwards from 10. Write down the numbers you say in your jotter.

| 10 | ? | ? | ? | ? | ? | ? | ? | ? | ? |

Blast off!

We can use the 1 to 9 pattern to help us count in tens.

Let's learn

Count forwards and backwards in tens with your teacher.

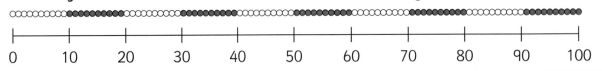

0   10   20   30   40   50   60   70   80   90   100

Let's practise

1) Count in tens. How many dots? Write the answer in your jotter.

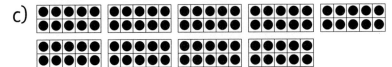

a)

b)

c)

2) Fill in the missing numbers on the caterpillar's body.

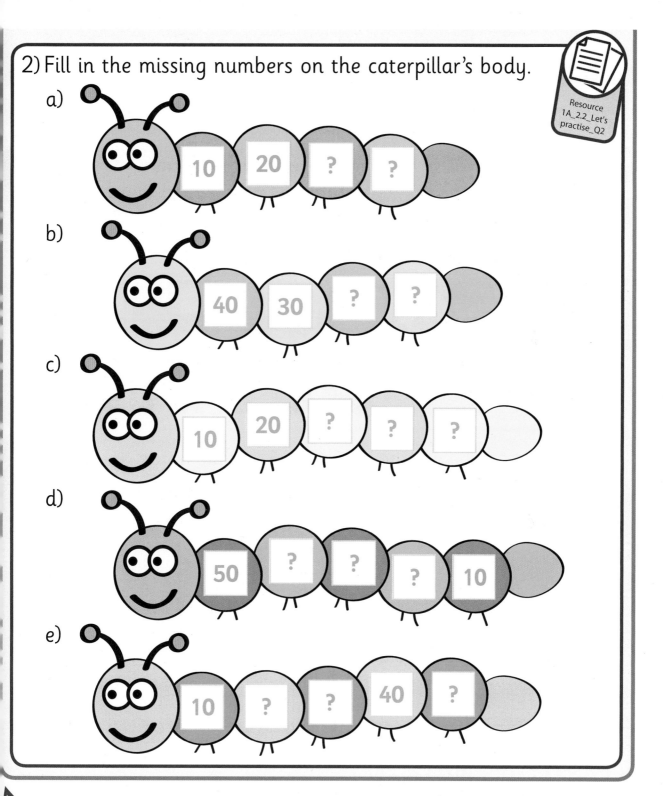

a) 10 20 ? ?

b) 40 30 ? ?

c) 10 20 ? ? ?

d) 50 ? ? ? 10

e) 10 ? ? 40 ?

⭐ CHALLENGE!

Use cubes to make towers. Put 10 cubes in each tower.
Count in tens to count the cubes. How far can you go?

2.3 Reading and writing numbers (2)

We are learning to read and write the tens numbers in words.

Before we start

Write the missing numbers.

a)

? 40 50 60 ? ?

b)

50 ? ? ? 90 ?

Numbers can be written in numerals and in words.

Let's learn

Talk with your teacher about these number words.

10	ten
20	twenty
30	thirty
40	forty
50	fifty
60	sixty
70	seventy
80	eighty
90	ninety
100	one hundred

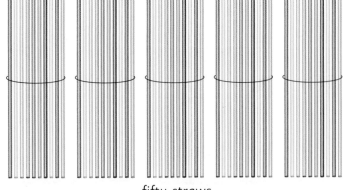

fifty straws

thirty apples

1) The children take the number **80** bus to the fair. Write this number in words in your jotter.

2)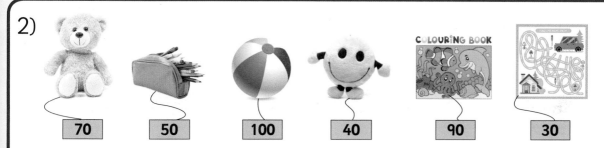

| 70 | 50 | 100 | 40 | 90 | 30 |

a) Nuria won prize number forty. What did she win?

b) Finlay won prize number ninety. What did he win?

c) Isla won prize number seventy. What did she win?

d) Amman won prize number fifty. What did he win?

3) a) What number is the ball? Write the number word.

b) What number is the maze? Write the number word.

CHALLENGE! ....

The number **27** can be written like this: **twenty-seven**.
Write these numbers in words in your jotter.
a) 24   b) 28   c) 26

2.4 Forward number sequences (1)

We are learning to read, write and say the numbers up to 100.

Before we start

Write the missing numbers in your jotter.

23	?	25	?	?	28	29	?

We can use the 0 to 9 pattern to help us remember the number order.

Let's learn

Talk about the patterns you see. Can you name the missing numbers? Read the numbers out loud in order.

1	2	3	4	5	6	7	8	9	10
11	12	13	14	15	16	17	18	19	?
21	?	23	?	?	?	27	28	29	30
31	32	?	34	35	36	37	38	39	40
41	42	?	44	?	46	47	48	49	50
51	52	53	54	55	?	?	58	59	60
61	62	63	64	?	66	?	68	?	70
71	?	?	?	?	76	77	78	79	80
81	82	83	84	85	86	87	88	?	?
?	92	93	94	95	96	97	?	99	100

1) What are the missing numbers? Write them in your jotter.

Resource 1A_2.4_Let's practise_Q1

a)
61 | ? | 63 | ? | 65 | ? | ? | ?

b)
43 | ? | ? | ? | 47 | 48 | ? | ?

c)
30 | 31 | ? | ? | 34 | ? | ? | ?

d)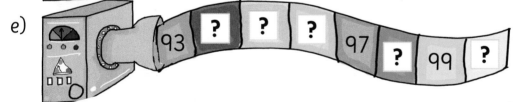
82 | ? | ? | 85 | ? | 87 | ? | ?

e)
93 | ? | ? | ? | 97 | ? | 99 | ?

CHALLENGE! ....

Write the missing bus seat numbers.

Resource 1A_2.4_Challenge

33 | | | 36 | | 38 | 39

| | | | 44 | 45 | | 47

2.5 Forward number sequences (2)

We are learning to count forwards in ones from any number up to 100.

Before we start

Finlay went to see a show. Write the missing seat numbers in your jotter.

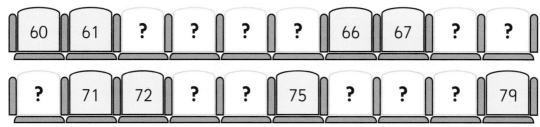

| 60 | 61 | ? | ? | ? | ? | 66 | 67 | ? | ? |

| ? | 71 | 72 | ? | ? | 75 | ? | ? | ? | 79 |

We can use the 1 to 9 pattern to help us read, write and say numbers in the correct order.

Let's learn

Talk about the snakes and ladders board.

Can you name the missing numbers?

Read the numbers out loud in order.

Resource
1A_2.5_Let's
learn

100	99			96			93	92	91
81	82	83	84		86			89	90
80	79			76	75		73		71
	62		64				68	69	
60	59	58			55			52	51
		43	44					49	50
40			37		35	34	33	32	
21			24			27			
20		18		16		14	13	12	11
1	2	3					8		10

1) The frog hops from stone to stone. Write the numbers he lands on then read them with a partner.

Resource 1A_
2.5_Let's
practise_Q1

a)
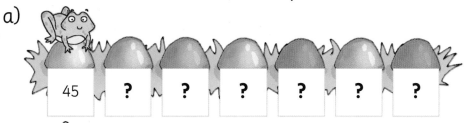

| 45 | ? | ? | ? | ? | ? | ? |

b)
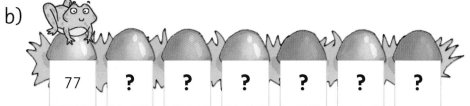

| 77 | ? | ? | ? | ? | ? | ? |

c)
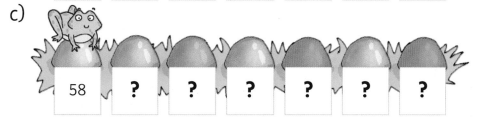

| 58 | ? | ? | ? | ? | ? | ? |

d)
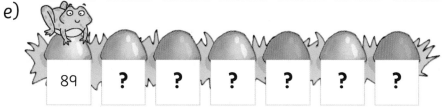

| 94 | ? | ? | ? | ? | ? | ? |

e)
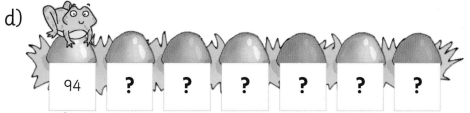

| 89 | ? | ? | ? | ? | ? | ? |

2) Jump along the number tracks.
Say and write the number you land on each time.
a) Start on 66. Take three jumps.
b) Start on 62. Take six jumps.

Resource
1A_2.5_Let's
practise_Q2

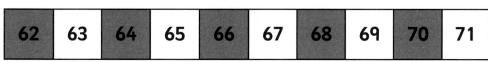

| 62 | 63 | 64 | 65 | 66 | 67 | 68 | 69 | 70 | 71 |

c) Start on 40. Take four jumps.

d) Start on 36. Take five jumps.

35	36	37	38	39	40	41	42	43	44

e) Start on 94. Take three jumps.

f) Start on 98. Take one jump.

90	91	92	93	94	95	96	97	98	99

g) Start on 48. Take seven jumps.

h) Start on 51. Take four jumps.

48	49	50	51	52	53	54	55	56	57

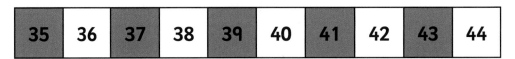

CHALLENGE!

a) How many jumps from 39 to 41?

b) How many jumps from 64 to 69?

c) How many jumps from 51 to 54?

d) How many jumps from 97 to 98?

2.6 Backward number sequences (1)

We are learning to count backwards.

Before we start

The boys line up in age order from oldest to youngest.

a) Write the missing ages in your jotter.

b) How old is the boy in red? Write the answer in words.

? 12 ? 10 ? 8 ? ? 5 4

We can use the 10 to 1 pattern to help us count backwards.

Let's learn

Talk about the patterns you see.

Can you name the missing numbers?

Read the numbers out loud in order from 100 to 1.

100	99	98	97	96	95	94	93	92	91
90	89	?	87	86	85	84	83	82	81
80	79	78	77	76	75	74	?	?	71
70	69	68	67	?	65	64	63	62	61
?	59	58	57	56	55	54	53	52	51
50	49	48	?	46	45	44	43	42	41
40	39	38	37	36	35	34	33	32	?
30	?	28	27	26	25	24	23	22	21
20	19	18	17	16	?	14	13	12	11
10	9	8	7	6	5	4	3	2	1

1) Write the numbers on the staircase steps as you count backwards in ones.

Resource 1A_2.6_Let's practise_Q1

a) 48   b) 59   c) 70   d) 89   e) 98
  47   58   69   88   97
    46   57   68   87   96

2) The ladybird lands on each leaf. Count backwards and write all the numbers she lands on in your jotter.

a) ? ? ? ? ? ?   77

b) ? ? ? ? ? ?   49

c) ? ? ? ? ? ?   100

d) ? ? ? ? ? ?   38

e) ? ? ? ? ? ?   96

CHALLENGE!

Count backwards in ones to complete the missing ticket numbers.

a) 31 30

b) 44 43

c) 72

d) 100

2.7 Backward number sequences (2)

We are learning to count backwards in 1s from any number between 0 and 50.

Before we start

Write the missing numbers.

14	13	?	?	10	?	8	?

We can use the 10 to 1 pattern to help us count backwards.

Let's learn

Amman and Isla are making a game. Talk about the board. Follow the arrows. Can you work out the missing numbers?

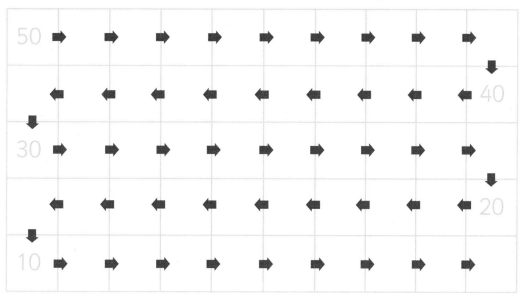

1) Count backwards and fill in the missing numbers.

a)

b)

c)

d)

e)

Resource 1A_2.7_Let's practise_Q1

CHALLENGE!

You will need a 100 square.
Say and write the number you land on each time.

a) Start on 62. Take three jumps backwards.
b) Start on 91. Take five jumps backwards.
c) Start on 75. Take eight jumps backwards.

100 square

2.8 Number before, after and in-between

> We are learning to say the number that comes before, after or in-between.

Before we start

Nuria has 31 number stickers to put in her album.

a) Draw the missing stickers and write the missing numbers in your jotter.

b) What number sticker is behind the star?

?	2	?	?	5	6	
?	8	?	10	11	12	
13	?	?	16	?	18	
?	?	21	22	23	?	
?	26	?	☆	29	?	31

> We count forwards to find the number after.
> We count backwards to find the number before.

The children draw a number line from 30 to 50.

| | | | | | | | | | | | | | | | | | | | | |
30 31 32 33 34 35 36 37 38 39 40 41 42 43 44 45 46 47 48 49 50

The number after 40 is 41.

The number before 40 is 39.

40 is in-between 39 and 41.

38, 39, 40 and 41 all come between 37 and 42.

Let's practise

1) Write the **number after**

Number lines

a) 37   b) 49   c) 83   d) 50   e) 99

f) 66   g) 29   h) 85   i) 70   j) 69

2) Write the **number before**

a) 60   b) 59   c) 91   d) 80   e) 39

f) 43   g) 28   h) 71   i) 90   j) 31

3) Say and write the **number in-between:**

a) | 69 | ? | 71 |

b) | 48 | ? | 50 |

c) | 79 | ? | 81 |

d) | 29 | ? | 31 |

★ **CHALLENGE!** ....

Write the numbers missing on these zoo tickets in your jotter.

a)

79	78	77	76	75
?	73	72	71	?
?	?	?	66	65
?	?	62	?	60

b) Write down in your jotter all the numbers that come **in-between** 66 and 73.

2.9 Counting in tens

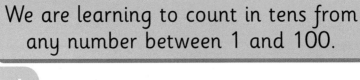

We are learning to count in tens from any number between 1 and 100.

Before we start

Write the missing numbers in your jotter.

| 10 | 20 | ? | ? | ? | ? | ? | ? | ? | ? |

We can count on in tens from any number.

Let's learn

Say Amman and Isla's numbers out loud.

1	2	3	4	5	6	7	8	9	10
11	12	13	14	15	16	17	18	19	20
21	22	23	24	25	26	27	28	29	30
31	32	33	34	35	36	37	38	39	40
41	42	43	44	45	46	47	48	49	50
51	52	53	54	55	56	57	58	59	60
61	62	63	64	65	66	67	68	69	70
71	72	73	74	75	76	77	78	79	80
81	82	83	84	85	86	87	88	89	90
91	92	93	94	95	96	97	98	99	100

Count on in tens from 3 to 93.

Count back in tens from 97 to 7.

1) Count on in tens. Write the missing numbers in your jotter.

a)

8 ? ? ? ? ? ? ? ? 98

b)

2 ? ? ? ? ? ? ? ? 92

c)

6 ? ? ? ? ? ? ? ? 96

2) Count back in tens. Write the missing numbers in your jotter.

a)

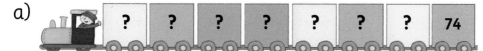
? ? ? ? ? ? ? 74

b)

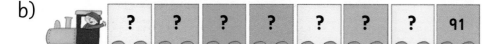
? ? ? ? ? ? ? 91

c)

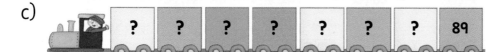
? ? ? ? ? ? ? 89

CHALLENGE!

Can you find the missing numbers?
Tell your teacher what they are.

Use a 100 square
to help.

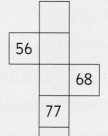

	34		56		
	44	45			68
53				77	

100 square

2.10 Base ten for teen numbers

> We are learning to find *how many* by building on ten.

Before we start

Find the missing numbers. $6 = 5$ and $\boxed{1}$ more.

Draw a bead string to match each number sentence.

a) $6 = 5$ and $\boxed{}$ more

b) $7 = 5$ and $\boxed{?}$ more

c) $9 = 5$ and $\boxed{?}$ more

d) $5 = 5$ and $\boxed{?}$ more

> We can count on from ten to find how many altogether.

Let's learn

Nuria shows that **10** and 4 makes **14** in different ways.

Bead string

Ten frames

Straws

Let's practise

1) Count on from ten. How many beads?

a)
b)
c)
d)

2) Copy and complete.

a) 10 and ? makes 15

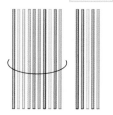

b) 10 and 3 makes ?

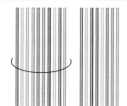

c) 10 and ? makes ?

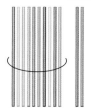

d) ? and ? makes ?

3) Both models show that **14** is the same as **10** and 4.

Ten frames

Use ten frames to model these number sentences.
Can you find more than one way?

a) **17** is the same as **10** and 7   b) **13** is the same as **10** and 3
c) **12** is the same as **10** and 2   d) **18** is the same as **10** and 8

⭐ CHALLENGE! ···

a) How many pencils?

10 Pencils

b) How many marbles?

10

2.11 Counting in tens and ones

We are learning to count in tens and ones to find how many.

Before we start

How many pencils?

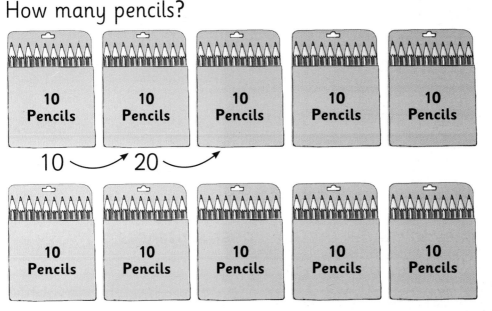

It helps to make groups of ten when we have lots of things to count.

Let's learn

Amman counts a pile of straws. He makes bundles of ten and has some straws left over. Amman counts the straws like this:

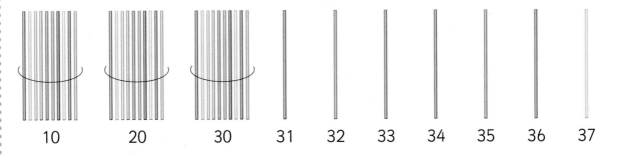

10   20   30   31   32   33   34   35   36   37

Nuria finds how many spots there are altogether. She knows the last ten frame has nine spots because there is one empty box. Nuria counts the spots like this:

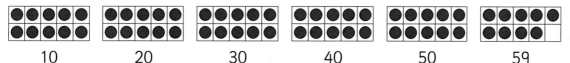

| 10 | 20 | 30 | 40 | 50 | 59 |

Let's practise

1) How many straws?

a)

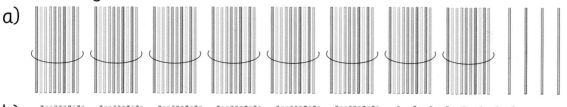

b)

2) How many dots?

a)

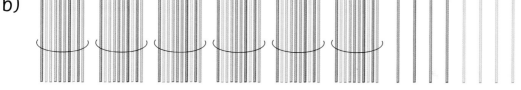

b)

3) How many beads?

a)

b)

CHALLENGE!

How much money does Nuria have?

2.12 Counting in twos and fives

> We are learning to count in twos and fives.

Before we start

How many sweets? Talk to a partner about how you worked it out.

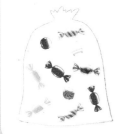    

> We know how to count in tens and ones. We can count things in twos and fives as well.

Let's learn

Count out loud in twos. How many dots altogether?

| 2 | 4 | 6 | 8 | 10 | 12 | 14 | 16 | 18 | 20 |

Count out loud in fives. How many dots altogether?

| 5 | 10 | 15 | 20 | 25 | 30 | 35 | 40 | 45 | 50 |

1) Count in twos to find how many. For example:

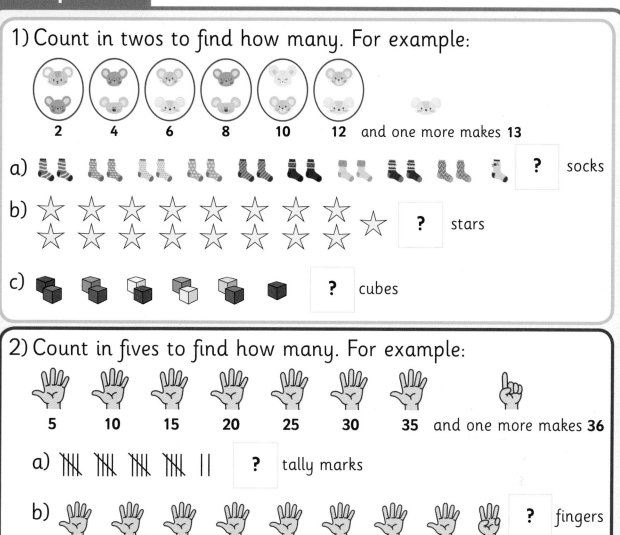

2   4   6   8   10   12   and one more makes **13**

a) ? socks

b) ? stars

c) ? cubes

2) Count in fives to find how many. For example:

5   10   15   20   25   30   35   and one more makes **36**

a) ? tally marks

b) ? fingers

c) ? cubes

⭐ CHALLENGE! ·

a) Start at 16. Keep counting in twos. How far can you go?
   16, 18, 20...

b) Start at 40. Keep counting in fives. How far can you go?
   40, 45, 50...

2.13 Ordinal number

> We are learning to describe the order of things.

Before we start

Isla is watching a car race.

a) Which car is in first place?
Write the car number in words.

b) Which car is in second place?
Write the car number in words.

c) Which car is in third place?
Write the car number in words.

> We can use words and symbols to describe the order of people, objects or events.

Let's learn

| first | second | third | fourth | fifth | sixth | seventh | eighth | ninth | tenth |
| 1st | 2nd | 3rd | 4th | 5th | 6th | 7th | 8th | 9th | 10th |

1) The pear is **first**. The apple is **tenth**. Write a word to describe the position of the:

   a) banana   b) lemon   c) carrot

   d) orange   e) pumpkin   f) tomato

2) The black pencil is **4th** in the line. Write symbols to describe the position of these pencils:

   a) red   b) pink   c) green

   d) purple   e) yellow   f) white

CHALLENGE!

Finlay's mum came **eighteenth** in a go-cart race. We can write eighteenth like this: **18th**.

Write down in your jotter where each driver finished. One has been done for you.

Finlay's mum → eighteenth → 18th

Dan	Pam	Jim	Tom
thirty-second	twenty-third	twenty-first	eleventh

3.1 Addition bonds to 10

We are learning to find addition partners.

Before we start

Which is the odd one out? Why?

| 7 + 3 | | 9 + 0 | | 6 + 4 |

| 5 + 5 | | 8 + 2 |

Every addition fact has a partner.

Let's learn

 $5 + 3 = 8$

 $3 + 5 = 8$

 $6 + 3 = 9$

$9 + 1 = 10$

$1 + 9 = 10$

 $3 + 6 = 9$

1) Write two addition facts for each picture in your jotter.

a)

b)

c)

d)

2) Write two addition facts for each domino in your jotter.

a)   b)   c)

d)   e)   f)

CHALLENGE! ....

You will need cards like these:

| 0 | 1 | 2 | 3 | 4 | 5 | 6 | 7 | 8 | 9 | 10 |

Number cards

Spread the cards out and turn them over.

Take turns to pick two cards.

If your cards make 10 you can keep them. If not, turn them back over.

The player who picks up the most cards is the winner.

3.2 Fact families within 10

We are learning to make fact families.

Before we start

Write a number sentence in your jotter to match each picture.

a)

? + ? = ?

b)

? – ? = ?

A fact family is two additions and two subtractions made with the same whole and parts.

Let's learn

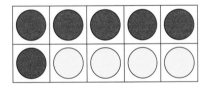   6 + 4 = 10

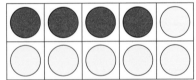   4 + 6 = 10

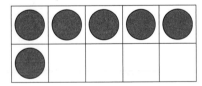   10 – 4 = 6

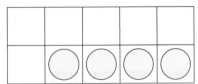   10 – 6 = 4

1) Write a fact family for each five frame in your jotter.

a)

b)

2) Write a fact family for each ten frame in your jotter.

a)

b)

c)

d)

CHALLENGE!

Isla writes a fact family to match this model.

3 + 4 = 7   4 + 3 = 7

7 − 3 = 4   7 − 4 = 3

Choose a number from 11 to 20.

Make your number with cubes of two different colours.

Ask a partner to write a fact family for your model.

3.3 Doubles to 20

We are learning to double 6, 7, 8, 9 and 10.

Before we start

Model each double on a ten frame. Write a number sentence to match each model.

Double 3 is 6
$3 + 3 = 6$

a) Double 2   b) Double 5   c) Double 4   d) Double 1

Double means to have **two lots** of the same number.

Let's learn

Double 6 is 12.

$6 + 6 = 12$

Double 3 is 6.

$3 + 3 = 6$

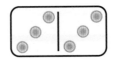

1) Write two different ways. One has been done for you.

Double 5 is 10
5 + 5 = 10

a)

b)

c)

d)

2) a) Double [**?**] = 12   b) [**?**] 10 = 20

c) Double [**?**] = 18   d) Double 7 = [**?**]

e) Double [**?**] = 16   f) 14 = double [**?**]

CHALLENGE!

Double 1 = 2

Double 2 = 4

Double 3 = 6

Double 4 = 8

Double 5 = 10

I know these double facts.

1) Help Finlay use the facts he knows to work out:

  a) double 10   b) double 20   c) double 30
  d) double 40   e) double 50   f) double 60

2) What other doubles can you work out?

3.4 Ten plus facts

> We are learning to add a one-digit number to 10.

Before we start

Copy and complete in your jotter.

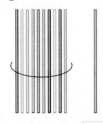

a) 11 is 10 and [?]   b) 13 is [?] and [?]   c) [?] is [?] and [?]

> We can use what we know about tens and ones to help us add numbers together.

Let's learn

The model shows that 7 + 7 = 14.

It also shows that **10 + 4 = 14**.

These ten frames show that **10 + 6 = 16**.

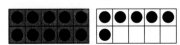

Talk about what you see.

What will 4 + 10 make? What about 6 + 10?

1) 10 red and 2 white makes 12 beads altogether. **10 + 2 = 12**

Write an addition number sentence in your jotter to match each bead string.

a) ⬤⬤⬤⬤⬤⬤⬤⬤⬤⬤◯◯◯◯◯

b) ⬤⬤⬤⬤⬤⬤⬤⬤⬤⬤◯◯◯◯◯◯◯

c) ⬤⬤⬤⬤⬤⬤⬤⬤⬤⬤⬤◯

d) ⬤⬤⬤⬤⬤⬤⬤⬤⬤◯◯◯

2) This model also shows that **10 + 2 = 12**

Write an addition number sentence to match each model.

a)   b)   c)   d)

3) Write a 'ten plus' fact for each pair of ten frames.

a)

b)

c)

d)

CHALLENGE!

Which model is the odd one out? Why?

a)

b)

c)

d)

e) ⬤⬤⬤⬤⬤⬤⬤⬤⬤⬤◯◯◯◯◯◯◯

f)

3.5 Five plus facts

We are learning to add 6, 7, 8 and 9 to the number 5.

Before we start

Match each pair of numbers that make 10.

When adding it helps to look for numbers that make 10.

Let's learn

Amman wants to add 6 and 5.

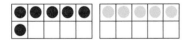

He moves four yellow counters over to make 10.

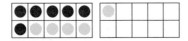

6 add 5 is the same as 10 add 1.   10 + 1 = 11 so 6 + 5 = 11.

1) Count, then model each picture with ten frames. Move counters to make 10. Write two addition facts to show what you did. Amman has done one for you.

Ten frames

 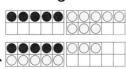

Move counters to make 10. 5 + 8 is the same as 10 + 3.

$5 + 8 = 13$ and $10 + 3 = 13$

a)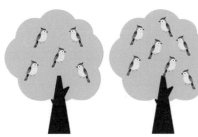

b)

c)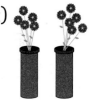

2) Copy and complete these additions. Use ten frames to help.

a) $7 + 5 =$?   b) $5 + 8 =$?

c) $9 + 5 =$?   d) $6 + 5 =$?

e) $5 + 7 =$?   f) $8 + 5 =$?

CHALLENGE!

a) Write a fact family to match the dice picture.

? + ? = ?   ? + ? = ?

? − ? = ?   ? − ? = ?

b) Roll a pair of dice. Write a fact family for the numbers rolled.

3.6 Addition bonds to 20

> We are learning to add two numbers by making 10.

Before we start

What addition could each model show?

a)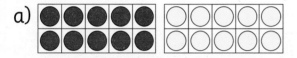

b)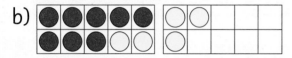

> A double ten frame can help with addition.

Let's learn

Let's add 4 + 9 on a blank double ten frame.

Put four dots on the top row and nine dots on the bottom row.

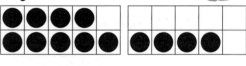

Move one dot from the bottom row up to the top row to make 10.

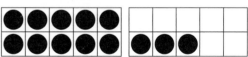

4 + 9 = 13

> I see 5 on the top and 8 on the bottom.
> 5 + 8 = 13

> I see 10 on the left and 3 on the right.
> 10 + 3 = 13

1) Make a double ten frame. Swap **one dot** to make 10.
   Copy the addition and write down the answer.

Ten frames

a)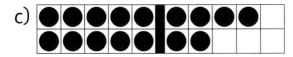

9 + 3 =

b)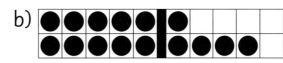

6 + 9 =

c)

9 + 7 =

d)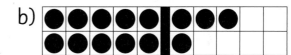

8 + 9 =

2) Make a double ten frame. Swap **two dots** to make 10.
   Copy the addition and write down the answer.

a)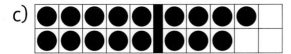

3 + 8 =

b)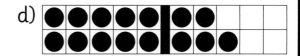

8 + 6 =

c)

9 + 8 =

d) 7 + 8 =

CHALLENGE! ....

Tell a partner how you would add these numbers
together. Show them on a double ten frame.

| 6 | 7 |

3.7 Subtraction within 20

We are learning to use partitioning to subtract.

Before we start

Find four different pairs of numbers that total 9.

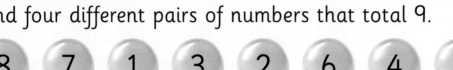

8 7 1 3 2 6 4 5

When we **partition** a number we break it into parts.

Let's learn

Let's find 12 – 7.

7 can be partitioned into 2 and 5.

12 beads subtract 2 beads leaves 10 beads.

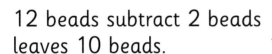

10 beads subtract 5 beads leaves 5 beads.

12 subtract 7 equals 5   12 – 7 = 5

1) Try these subtractions. Make the first number on a bead string then **partition** the second number to find the answer. One has been done for you.

a) $14 - 8 = $ **?**

Partition 8 into 4 and 4.

$14 - \mathbf{4} = 10$ and $10 - \mathbf{4} = 6$ so $14 - \mathbf{8} = 6$

b) $11 - 3 = $ **?**

c) $13 - 5 = $ **?**

d) $18 - 9 = $ **?**

e) $15 - 7 = $ **?**

f) $14 - 6 = $ **?**

g) $11 - 2 = $ **?**

CHALLENGE!

Which subtraction is the odd one out? Explain.

$11 - 5$ $15 - 9$ $12 - 6$ $17 - 8$ $13 - 7$

3.8 Counting on and counting back

We are learning to count on and back to solve addition and subtraction problems.

Before we start

Amman must count on 5. He thinks he will land on 6.

| 1 | | 3 | 4 | 5 | 6 | 7 | 8 | 9 | 10 |

Isla must also count on 5. She thinks she will land on 10.

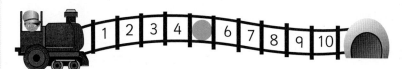

| 1 | 2 | 3 | 4 | | 6 | 7 | 8 | 9 | 10 |

Who is correct? Explain.

We can count on to add. We can count back to subtract.

Let's learn

This frog is counting on.

add 3

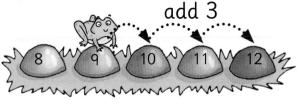

| 8 | 9 | 10 | 11 | 12 |

| 9 | + | 3 | = | 12 |

This ladybird is counting back.

take away 2

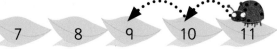

| 7 | 8 | 9 | 10 | 11 |

| 11 | − | 2 | = | 9 |

1) Count on to add. Write the number sentence in your jotter.

a)

add 3

? + ? = ?

b)

add 3

? + ? = ?

c)

plus 4

? + ? = ?

d)

add 0

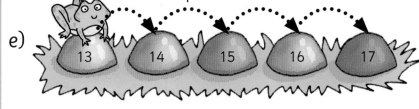

? + ? = ?

e)

plus 4

? + ? = ?

f)

plus 2

? + ? = ?

2) Count back to subtract.
   Write the number sentence in your jotter.

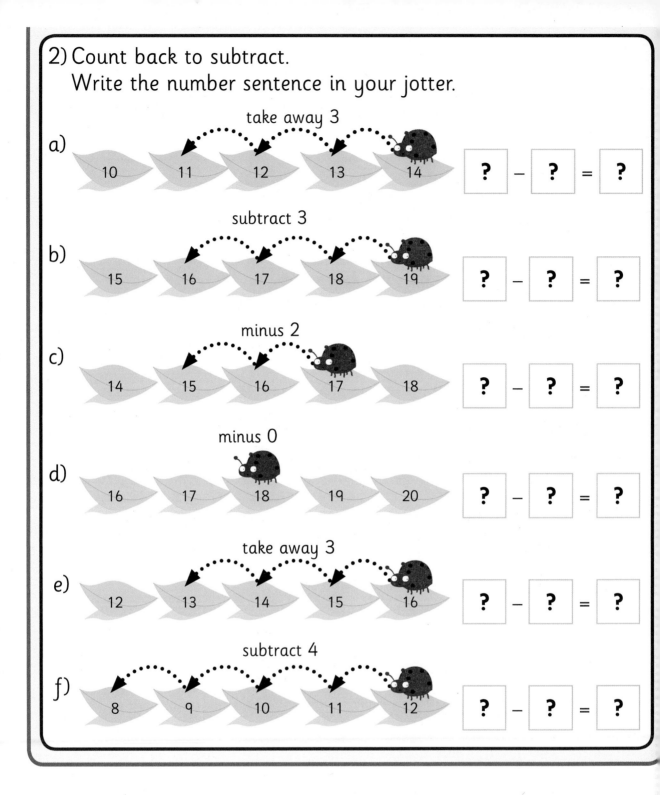

a) take away 3

10 11 12 13 14

? − ? = ?

b) subtract 3

15 16 17 18 19

? − ? = ?

c) minus 2

14 15 16 17 18

? − ? = ?

d) minus 0

16 17 18 19 20

? − ? = ?

e) take away 3

12 13 14 15 16

? − ? = ?

f) subtract 4

8 9 10 11 12

? − ? = ?

1) Write an addition number sentence in your jotter to match each frog's journey.

   a)   add 5

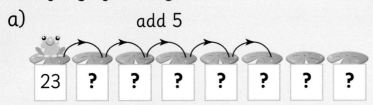

| 23 | ? | ? | ? | ? | ? | ? | ? |

   b)   plus 6

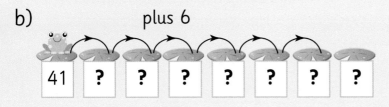

| 41 | ? | ? | ? | ? | ? | ? | ? |

   c)   plus 5

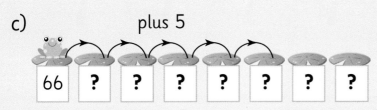

| 66 | ? | ? | ? | ? | ? | ? | ? |

2) Write a subtraction number sentence in your jotter to match each ladybird's journey.

   a)   subtract 5

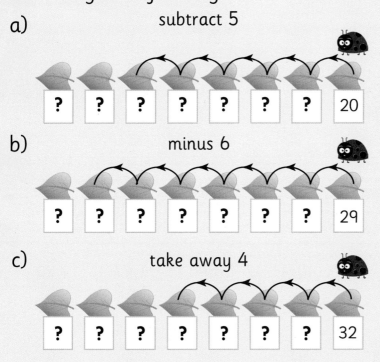

| ? | ? | ? | ? | ? | ? | ? | 20 |

   b)   minus 6

| ? | ? | ? | ? | ? | ? | ? | 29 |

   c)   take away 4

| ? | ? | ? | ? | ? | ? | ? | 32 |

3.9 Adding two one-digit numbers using a number line

We are learning to add using a number line.

Before we start

Write an addition number sentence to match the picture.

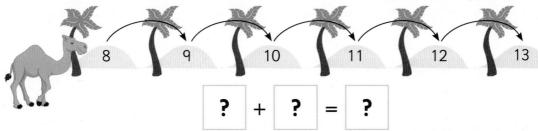

| ? | + | ? | = | ? |

We can jump forwards on a number line to help us with addition.

Let's learn

Talk about how the girls have solved this addition problem. **8 + 6 =** 14

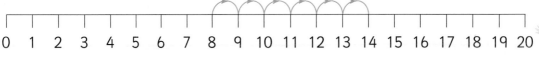

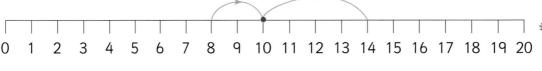

1) Use the number lines to complete these additions.

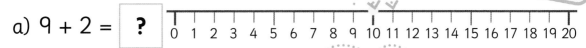

a) $9 + 2 =$ [?] | 0 1 2 3 4 5 6 7 8 9 10 11 12 13 14 15 16 17 18 19 20

b) $7 + 5 =$ [?] | 0 1 2 3 4 5 6 7 8 9 10 11 12 13 14 15 16 17 18 19 20

c) $5 + 7 =$ [?] | 0 1 2 3 4 5 6 7 8 9 10 11 12 13 14 15 16 17 18 19 20

d) $8 + 4 =$ [?] | 0 1 2 3 4 5 6 7 8 9 10 11 12 13 14 15 16 17 18 19 20

e) $9 + 8 =$ [?] | 0 1 2 3 4 5 6 7 8 9 10 11 12 13 14 15 16 17 18 19 20

f) $8 + 3 =$ [?] | 0 1 2 3 4 5 6 7 8 9 10 11 12 13 14 15 16 17 18 19 20

g) $4 + 9 =$ [?] | 0 1 2 3 4 5 6 7 8 9 10 11 12 13 14 15 16 17 18 19 20

CHALLENGE! ....

Jump along a number line to find the answers to these additions. Draw the jumps you take.

a) $9 + 4 =$ [?] | 0 1 2 3 4 5 6 7 8 9 10 11 12 13 14 15 16 17 18 19 20

b) $3 + 8 =$ [?] | 0 1 2 3 4 5 6 7 8 9 10 11 12 13 14 15 16 17 18 19 20

c) [?] $= 7 + 6$ | 0 1 2 3 4 5 6 7 8 9 10 11 12 13 14 15 16 17 18 19 20

d) [?] $= 5 + 9$ | 0 1 2 3 4 5 6 7 8 9 10 11 12 13 14 15 16 17 18 19 20

3.10 Subtracting using a number line

We are learning to subtract using a number line.

Before we start

Write a subtraction number sentence to match the picture.

15 take away 3

| ? | – | ? | = | ? |

We can jump backwards on a number line to help us with subtraction.

Let's learn

Talk about how the boys have solved this subtraction problem. **14 – 7 =** 7

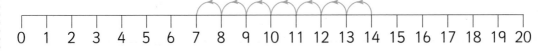

0 1 2 3 4 5 6 7 8 9 10 11 12 13 14 15 16 17 18 19 20

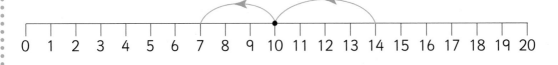

0 1 2 3 4 5 6 7 8 9 10 11 12 13 14 15 16 17 18 19 20

1) Use the number lines to complete these subtractions.

Number lines

a) $12 - 8 =$?

0 1 2 3 4 5 6 7 8 9 10 11 12 13 14 15 16 17 18 19 20

b) $11 - 6 =$?

0 1 2 3 4 5 6 7 8 9 10 11 12 13 14 15 16 17 18 19 20

c) $18 - 9 =$?

0 1 2 3 4 5 6 7 8 9 10 11 12 13 14 15 16 17 18 19 20

d) $16 - 8 =$?

0 1 2 3 4 5 6 7 8 9 10 11 12 13 14 15 16 17 18 19 20

e) $13 - 6 =$?

0 1 2 3 4 5 6 7 8 9 10 11 12 13 14 15 16 17 18 19 20

f) $15 - 6 =$?

0 1 2 3 4 5 6 7 8 9 10 11 12 13 14 15 16 17 18 19 20

g) $13 - 9 =$?

0 1 2 3 4 5 6 7 8 9 10 11 12 13 14 15 16 17 18 19 20

CHALLENGE! ....

Jump along a number line to find the answers to these subtractions. Draw the jumps you take.

Number lines

a) $13 - 4 =$?

0 1 2 3 4 5 6 7 8 9 10 11 12 13 14 15 16 17 18 19 20

b) $11 - 8 =$?

0 1 2 3 4 5 6 7 8 9 10 11 12 13 14 15 16 17 18 19 20

c) ? $= 15 - 9$

0 1 2 3 4 5 6 7 8 9 10 11 12 13 14 15 16 17 18 19 20

d) ? $= 14 - 5$

0 1 2 3 4 5 6 7 8 9 10 11 12 13 14 15 16 17 18 19 20

3.11 Missing addend

We are learning to solve missing number problems.

Before we start

How many more cubes are needed to make each row 10 cubes long?

a) $5 + \boxed{?} = 10$

b) $8 + \boxed{?} = 10$

c) $9 + \boxed{?} = 10$

d) $3 + \boxed{?} = 10$

We can count on to solve missing number problems.

Let's learn

Talk about **how** Finlay and Nuria have solved this problem.

 and *some* blue counters under here **makes 13 altogether**

$9 \qquad + \qquad \boxed{?} \qquad = 13$

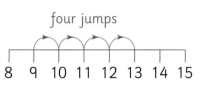

Let's practise

1) Find the missing number then write down the number sentence.

a)

and *some* blue counters under here makes 13 altogether

8 + **?** = 13

b)

and *some* blue counters under here makes 11 altogether

6 + **?** = 11

c)

and *some* blue counters under here makes 12 altogether

6 + **?** = 12

d)

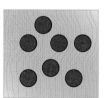

and *some* blue counters under here makes 13 altogether

7 + **?** = 13

2) a) How many more mice to make 11?

? + ? = 11

b) How many more cars to make 12?

? + ? = 12

c) How many more cats to make 11?

? + ? = 11

d) How many more pencils to make 12?

 ? + ? = 12

e) How many more flowers to make 14?

 ? + ? = 14

f) How many more bees to make 14?

 ? + ? = 14

★ CHALLENGE! ⋯⋯⋯⋯⋯⋯⋯⋯⋯⋯⋯⋯⋯⋯⋯⋯⋯

Show how you would work each answer out on a number line.

a) ? + 15 = 17   b) ? + 17 = 20

c) ? + 14 = 18   d) ? + 13 = 19

3.12 Missing subtrahend

> We are learning to solve missing number problems.

Before we start

a)

four dots under here

This is 17

What number is this dot?

b)

three cubes under here

This is 21

What number is this cube?

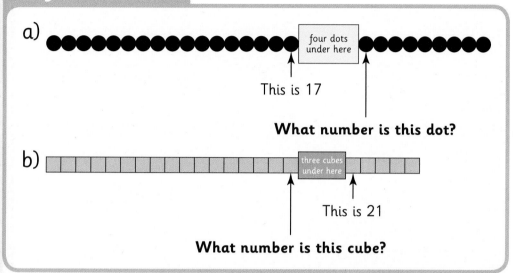

> Subtraction problems can be solved by counting on or counting back.

Let's learn

Miss Higgins counts out **11** cubes. The children close their eyes and Miss Higgins hides some of the cubes in a box.

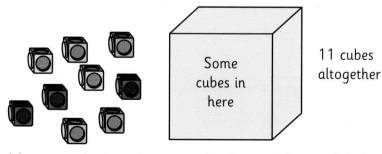

Some cubes in here

11 cubes altogether

How many cubes are left on the table?
How many cubes are in the box?

Isla counts back. She writes a number sentence to match her thinking.

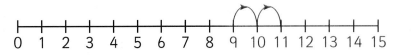

$$11 - \boxed{?} = 9$$

Amman counts on. He writes a number sentence to match his thinking.

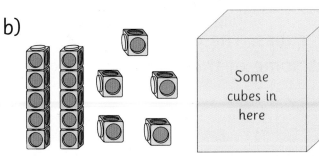

$$9 + \boxed{?} = 11$$

Talk about the different ways to solve this problem.

Let's practise

1) How many cubes in each box?

Write a number sentence to match how you worked it out.

a)

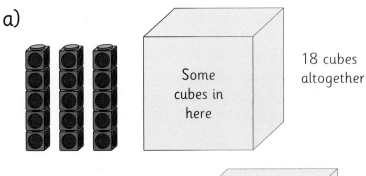

Some cubes in here

18 cubes altogether

b)

Some cubes in here

17 cubes altogether

2) Solve by counting on or counting back.
Write a number sentence to match your thinking.

a) **11** sweets altogether. How many sweets in the bag?

b) **12** counters altogether. How many counters in the cup?

c) **12** children altogether. How many children behind the wall?

d) **13** cows altogether. How many cows in the barn?

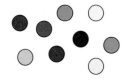

CHALLENGE! ....

Find the missing numbers

a) $9 + \boxed{?} = 15$

b) $\boxed{?} + 9 = 15$

c) $15 - \boxed{?} = 9$

d) $15 - \boxed{?} = 6$

3.13 Representing and solving word problems (1)

> We are learning to solve addition and subtraction story problems.

Before we start

a) Isla had three sweets. Nuria gave her five more.

How many sweets does Isla have now?

b) Finlay had 10 sweets. He gave three to Amman. How many sweets does Finlay have now?

> Objects and drawings can help us think about story problems.

Let's learn

Mrs Marr has two cats and nine fish.
How many pets does Mrs Marr have?

The children write down the number sentence $2 + 9 =$ ❓

Nuria uses a double ten frame and counters.

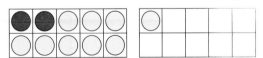

Finlay draws a number line.

Isla draws the pets and counts how many there are altogether.

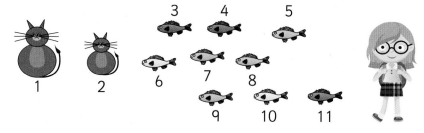

All of the children write down 2 + 9 = 11
Mrs Marr has 11 pets.

Let's practise

Write down the number sentence and the answer.

1) Isla had nine flowers. She picked nine more from the garden. How many does she have now?

2) There were 16 birds in the tree. Eight flew away. How many birds are there now?

3) At a party nine children wanted orange juice and eight wanted cola. How many children wanted a drink?

4) Nuria has 12 red beads and 8 blue beads. How many beads does she have?

CHALLENGE!

Finlay has 16 stars. Nine are silver and the rest are gold. How many gold stars does Finlay have?

3.14 Representing and solving word problems (2)

We are learning to think about the same story problem in different ways.

Before we start

How many more are needed to make 12?

a) | ? | pencils

b) | ? | toy cars

c) | ? | marbles

We can model the same problem in different ways.

Let's learn

I need 16 cakes for my party. I have nine cakes. How many more do I need to buy?

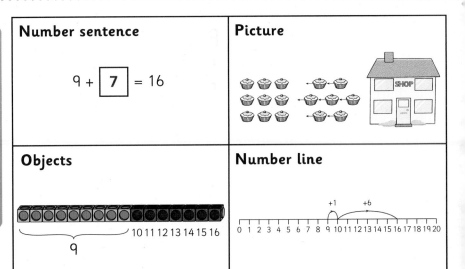

Number sentence	Picture
$9 + \boxed{7} = 16$	
Objects	**Number line**
9 10 11 12 13 14 15 16	+1 +6 0 1 2 3 4 5 6 7 8 9 10 11 12 13 14 15 16 17 18 19 20

Solve each problem. Write a number sentence to show your thinking.

1) Finlay had 12 stickers. Amman gave him some more. Now he has 15.
How many stickers did Amman give to Finlay?

2) There were some fish in Mr Smith's pond. He bought three more. Now he has 11.
How many fish did Mr Smith start with?

3) Isla had 11 bears. She gave her sister some. Now she has seven.
How many bears did Isla give away?

4) There were 14 cars in the car park. Five more came. How many cars are in the car park now?

5) Nuria went to the shop. She spent £10. When she got home, she had £2 left.
How much did Nuria have to start with?

CHALLENGE!

Make up a story problem for each number sentence. Ask a partner to work out the answers.

| 6 + 11 | 17 − 4 |

3.15 Representing and solving word problems (3)

We are learning to find the difference using a bar model.

Before we start

There are 11 children in a team. Five are girls and the rest are boys. How many boys are there?

A bar model can help us to compare two amounts.

Let's learn

Finlay has eight robots and three spaceships. How many more robots are there?

Objects

Bar model

8	
3	?

The difference between 8 and 3 is 5   3 + ⬚5⬚ = 8   8 – 3 = ⬚5⬚

Let's practise

Complete the questions like the example.
Use cubes or counters to help you model the story.

Blank bar models

1) Nuria has 10 green cubes and 7 blue cubes. How many more green cubes does she have?

2) Isla has 12 bears. Some are brown and nine are white.
How many brown bears does she have?

3) Amman has 20 counters.
10 are red and the rest are yellow.
How many yellow counters are there?

4) There are 11 girls and 7 boys.
If all the girls take a boy as a partner, how many boys won't have a partner?

CHALLENGE!

a) Write an addition number sentence and a subtraction number sentence to fit this bar model.

18	
10	8

b) Now write a story problem to fit one of your number sentences.

4.1 Making equal groups

> We are learning to make and describe equal-sized groups.

Before we start

What does this say? Can you get this many counters?

14

> We can use counters or other objects to help us make equal groups.

Let's learn

Finlay has lots of marbles to count. Isla helps him by putting them into groups of two.

Point to each group as you count them:
2 ... 4 ... 6 ... 8 ... 10 ... 12

He puts his cars into groups of three.

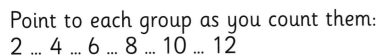

Help him count the cars: 3 ... 6 ... 9 ... 12 ... 15

Can you put objects into equal-sized groups now?

Let's practise

1) Count out 12 objects. Put them into equal groups of four. How many groups have you got?

2) Make three groups of five counters. How many counters do you have altogether?

3) Nuria has made eight cupcakes. She puts two on each plate. How many plates will she need? Use counters and put them into groups of two to help you.

4) Baha the butterfly wants to know how many caterpillars she has! Can you help her?

Draw the caterpillars on the leaves and count the total.

a) Two caterpillars on each leaf. How many caterpillars?

b) Three caterpillars on each leaf. How many caterpillars?

c) Four caterpillars on each leaf. How many caterpillars?

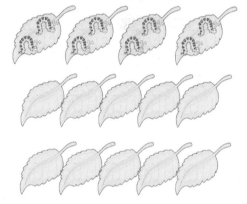

CHALLENGE!

a) Count out 20 cubes. Now build them into towers of five. How many towers do you have?

b) Now sort 20 cubes into towers of two. How many towers now?

c) Can you sort the cubes into equal-sized towers any other ways?

4.2 Using arrays

We are learning to make and draw arrays to represent equal groups of objects.

Before we start

Here are 18 counters.

Can you sort these into groups of three? How many groups do you have?

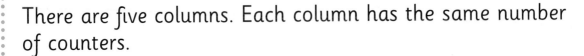

We can organise equal-sized groups of objects into rows and columns. This is called an **array**.

Let's learn

Amman has three groups of five counters. Let's organise them into an array.

There are three rows. Each row has the same number of counters.

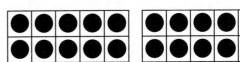
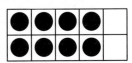

There are five columns. Each column has the same number of counters.

How many counters are there altogether? Let's count:
5 … 10 … 15.

1) Here are some arrays. Can you say how many rows and how many columns for each one?

a)   ● ● ●
     ● ● ●
     ● ● ●
     ● ● ●

   [?] rows   [?] columns

b)   ● ● ● ● ●
     ● ● ● ● ●

   [?] rows   [?] columns

c)   ● ● ● ● ● ●
     ● ● ● ● ● ●
     ● ● ● ● ● ●

   [?] rows   [?] columns

2) a) Count out eight counters. Sort them into rows of four to make an array. How many rows do you have?

b) Add another row of four to your array. How many rows do you have now?

c) Add one more row of four to your array. How many rows do you have now? How many counters are there now?

3) Sort these groups into rows and columns and draw an array for each one.

a) Five groups of two.
b) Three groups of four.
c) Five groups of ten.
d) Six groups of three.

> Use counters or cubes to help you.

CHALLENGE! ....

Count out 24 counters. How many different arrays could you make by organising the counters into different-sized rows and columns? Draw all the arrays you find!

4.3 Skip counting in twos

We are learning to skip count in twos.

Before we start

How many altogether? Can you write how many?

When we skip count in twos, we skip over a number.

Let's learn

If we start from 0, we can skip over the odd numbers to count in twos. These are the even numbers:

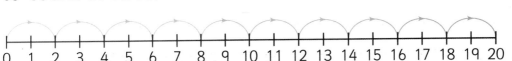

If we start from 1, we can skip over the even numbers to count in twos. These are the odd numbers:

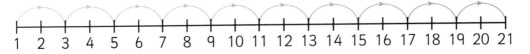

We can start from any number and skip count in twos:

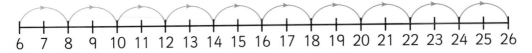

We can skip count backwards too:

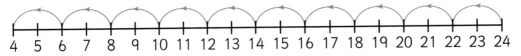

Let's practise

1) Work with a partner and skip count in twos starting from these numbers:

   a) 2   b) 1   c) 5   d) 4

   Write down if you are counting in even numbers or odd numbers for each question. Tell a partner how you know.

2) Skip count forwards in twos. How many animals are there in each group?

   a)       

   b)

   c)

3) Skip count backwards in twos. Count along the leaves, saying one number in the sequence for each leaf. Write down the missing numbers.

   18   14   10   8   2

CHALLENGE!

Ten frames

You will need blank ten frames for this question.
Take four ten frames and cover them with counters.

a) Skip count in twos to find out how many counters you have altogether.

b) Try adding one more ten frame. How many counters now?

4.4 Skip counting in tens

We are learning to skip count in tens.

Before we start

a) What is this number?

b) What number comes after this?

c) What number comes before it?

49

When we skip count in tens, we skip over numbers to just count the multiples of ten.

Let's learn

If we start from 0 and skip count in tens, we can see that all the numbers end in 0:

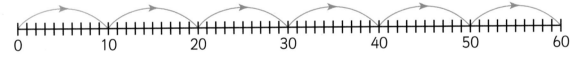

0   10   20   30   40   50   60

We can also use a hundred square. Start from 0 and count in tens. Place counters over all the tens. What do you notice?

1	2	3	4	5	6	7	8	9	10
11	12	13	14	15	16	17	18	19	20
21	22	23	24	25	26	27	28	29	30
31	32	33	34	35	36	37	38	39	40
41	42	43	44	45	46	47	48	49	50
51	52	53	54	55	56	57	58	59	60
61	62	63	64	65	66	67	68	69	70
71	72	73	74	75	76	77	78	79	80
81	82	83	84	85	86	87	88	89	90
91	92	93	94	95	96	97	98	99	100

We can skip count backwards too:

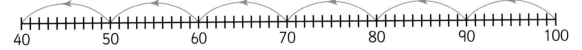

40   50   60   70   80   90   100

1) Work with a partner and take turns to skip count in tens out loud. What do you notice about all the numbers you say? You could use a number line or hundred square to help you.

a) Start at 20 and stop at 80.　　b) Start at 50 and stop at 100.

c) Start at 70 and stop at 10.　　d) Start at 100 and stop at 30.

2) There are 10 sweets in each bag. Skip count forwards in tens to find out how many sweets there are in total.

3) Skip count backwards in tens. Write down the missing numbers.

?　　**40**　　**?**　　**20**　　**?**

CHALLENGE!

Amman is skip counting along these stepping stones.

500　400　300

a) What number is he skip counting in?

b) Is he skip counting forwards or backwards?

c) What numbers do you think he would say for the last two stepping stones? Write them down in your jotter.

4 Number – multiplication and division

4.5 Skip counting in fives

We are learning to skip count in fives.

Before we start

Look at these numeral cards:

Can you point to the number 41? How do you know you're right? What about 14 – can you point to that one? Can you count from 14 to 41?

| 14 | 21 | 28 | 35 | 41 |

When we skip count in fives, we skip over numbers to just count the multiples of five.

Let's learn

If we start from 0 and skip count in fives, we can see that all the numbers end in 0 or 5:

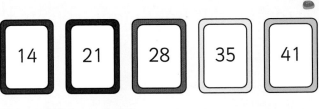

We can start from any number:

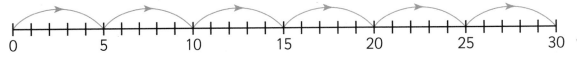

We can skip count backwards too:

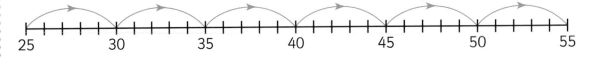

1) Work with a partner and take turns to skip count in fives out loud.

Number lines and 100 square

   a) Start at 15 and stop at 40.
   b) Start at 50 and stop at 10.
   c) Start at 70 and stop at 100.
   d) Start at 90 and stop at 35.

2) a)

   How many fingers?

   b)

   How many fingers?

   c)

   How many fingers?

   d)

   How many fingers?

3) Oops! Nuria has spilled some paint. Write down the numbers behind the splashes.

   a) 35, 40, ✽ 50, ✽, ✽ 65   b) 10, ✽ 20, 25, ✽ 35, ✽

   c) 50, ✽ 40, ✽ ✽, 25   d) ✽, 25, 20, ✽ ✽ ✽

CHALLENGE! ....

100 square

On a 100 square, skip count in twos and colour all those numbers in yellow. Skip count in fives and colour those numbers in green. Then skip count in tens and colour those numbers in red.

4.6 Multiplying using skip counting

> We are learning to use skip counting to count equal groups.

Before we start

Here are 30 counters.
Sort them into equal groups
of five. How many groups do you have?

> We can use skip counting to find the total number of groups.

Let's learn

Here is an array:

There are three rows of five.

We can use a number line
to help us skip count:

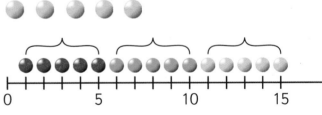

We skip count three times
and the total is 5 … 10 … 15.

We can record this problem like this: 3 groups of 5 = 15

Or like this: 5 + 5 + 5 = 15

Or we can use the **multiply** symbol like this: 3 × 5 = 15

1) Make an array to show these problems. Then skip count to work out how many altogether.

You can use a number line to help you.

a) Four groups of two.
b) Five groups of three.
c) Three rows of ten.
d) Six groups of five.
e) Seven rows of two.
f) Five groups of ten.

2) There are lots of rollercoasters at the funfair! How many people are on each rollercoaster? Skip count on the number line to find the total.

a) The red rollercoaster has four cars. Each car has 10 people in it. How many people?

```
0   5   10   15   20   25   30   35   40   45   50
```

b) The blue rollercoaster has nine cars. Each car has five people in it. How many people?

```
0   5   10   15   20   25   30   35   40   45   50
```

c) The green rollercoaster has 17 cars. Each car has two people in it. How many people?

```
0  2  4  6  8 10 12 14 16 18 20 22 24 26 28 30 32 34 36 38 40
```

CHALLENGE!

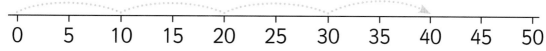

Try some multiplication sums.
Skip count on a number line to help you.

a) $7 \times 3 =$?

b) $8 \times 4 =$?

c) $3 \times 5 =$?

d) $7 \times 10 =$?

4.7 Sharing equally

> We are learning to divide collections by sharing equally.

Before we start

Isla has five pairs of socks. Work out how many socks she has.

> We can use objects to help us share into equal groups.

Let's learn

Finlay has six sweets.

He shares them with Amman and Nuria.

Use six counters. Share them equally between the three children.

They each get two sweets!

Six shared between three is two.

We can use the **divide** symbol like this: 6 ÷ 3 = 2

1) Isla is having a party. Share these things out equally.

Use counters or counting objects to help you.

   a) 12 cupcakes between 3 plates. How many on each plate?

   b) 25 sausage rolls between 5 plates. How many on each plate?

   c) 20 balloons between 10 party bags. How many in each bag?

   d) 30 crayons between 10 party bags. How many in each bag?

2) Share the objects between the children. How many will each child have? Use counters or counting objects to help you.

   a) Share 20 socks between 5 children.

   b) Share 12 cupcakes between 6 children.

   c) Share 35 crayons between 7 children.

   d) Share 40 biscuits between 8 children.

CHALLENGE!

Finlay's mother is cooking 24 mini pizzas for Finlay and his friends. There is an equal number of pizzas for each child.

How many children could there be, with how many pizzas for each child?

There is more than one correct answer!

4.8 Grouping

We are learning to divide collections by grouping them into equal sets.

Before we start

Nuria is counting in fives, what are the next three numbers?

5, 10, 15 ...

We can use counters or other objects to help us divide collections into equal groups.

Let's learn

Amman is visiting a farm. The hens have laid 18 eggs.

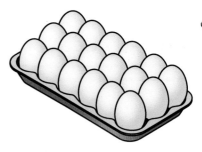

They are put into boxes of six. How many boxes will the farmer need?

We can divide them into groups of six to find out.

Count out 18 counters or cubes and sort them into groups of 6.

He needs three boxes!

We can use the **divide** symbol like this: 18 ÷ 6 = 3

1) The farmer is building some new pens for his animals.

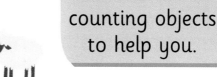
Use counters or counting objects to help you.

a) He has 20 sheep and he puts 5 sheep in each pen. How many pens will he need?

b) He has 10 cows and he puts 2 cows in each pen.

How many pens will he need?

c) He has 30 pigs and he puts 10 in each pen.

How many pens will he need?

2) Isla is sorting her baking into boxes. Use counters to work out how many boxes she will need.

a) 12 cupcakes into boxes of 3 each. How many boxes?

b) 25 biscuits into boxes of 5 each. How many boxes?

c) 16 muffins into boxes of 2 each. How many boxes?

d) 20 brownies into boxes of 4 each. How many boxes?

CHALLENGE!

Nuria's class is going on a school trip. Her teacher is sorting them into equal-sized groups. There are 40 children altogether. Work with a partner and write down all the different ways her teacher could group the children. You could use counting objects or draw pictures to help you work it out.

5.1 Make and identify halves

We are learning to make and identify halves.

Before we start

Which shapes show halves?

When we half something, we split it into two equal parts.

Let's learn

The boys cut their sandwiches in half. They cut them into two parts:

 Your two parts ARE NOT the same size. You HAVE NOT cut your sandwich in half.

Your two parts ARE the same size. You HAVE cut your sandwich in half.

Nuria makes halves in a different way to Amman:

   **is equal to**

These parts are both half of the sandwich.

There is exactly the same amount to eat in each part.

Halves **must be** the **same size**.

Halves **do not have to** be the **same shape**.

Let's practise

1) Have these shapes been spilt in half? YES or NO

a)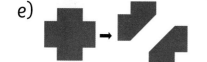

b)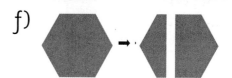

c)

d)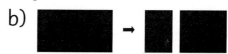

e)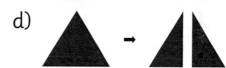

f)

2) Match the half shapes with the correct whole shape (the first one has been done for you):

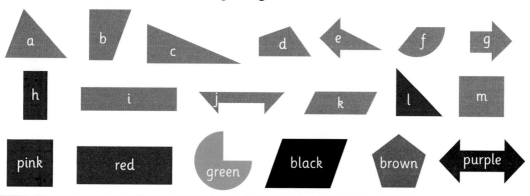

3) These patterns have been painted using half yellow and half red paint:

Which of the following patterns show exactly half red and half yellow?

a)

b)

c)

d)

e)

f)

⭐ **CHALLENGE!** ....

Amman is using squared paper to make a pattern showing halves:

Ask for squared paper.

Make a pattern using two different colours to show halves.

 →

Half my pattern is blue and half my pattern is green.

Squared paper

5.2 Make and identify quarters

We are learning to make and identify quarters.

Before we start

Which shapes have been split into four equal parts?

a)   b)   c) ▭ d) ◥◣◥◣

When we quarter something, we split it into four equal parts.

Let's learn

The girls have each been asked to cut a block of modelling clay into quarters. They cut them into four parts:

Nuria's four parts ARE NOT the same size. They are NOT quarters.

| one quarter | one quarter |
| one quarter | one quarter |

Isla's four parts ARE the same size. They are quarters.

Amman makes quarters in a different way to Isla:

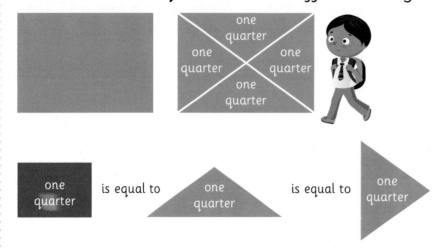

These parts are all a quarter of the clay.

There is exactly the same amount of clay in each of these parts.

Quarters **must be** the **same size**.

Quarters **do not have to** be the **same shape**.

Let's practise

1) Have these shapes been spilt into quarters? YES or NO

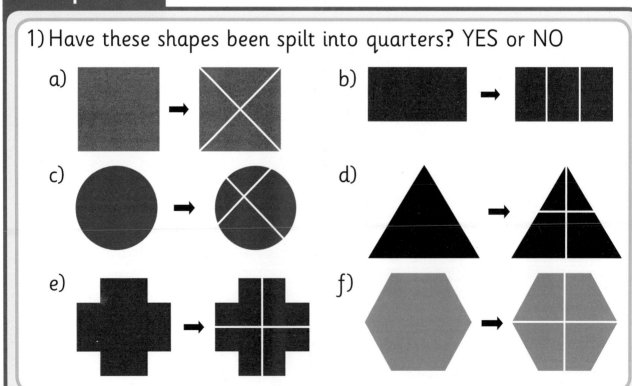

2) Match the quarter shapes with the correct whole shape (one has been done for you):

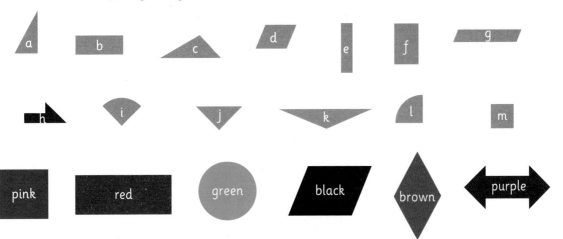

3) Which of the following patterns show quarters?

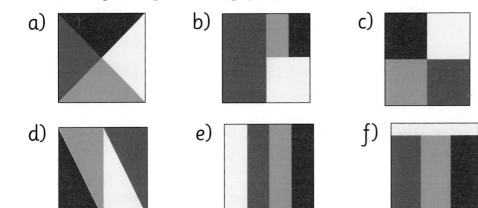

a)

b)

c)

d)

e)

f)

⭐ **CHALLENGE!** ....

Ask for different-shaped pieces of paper like these (you can use different shapes if you have them).

Fold or cut them to make quarters.

Can you make quarters with all of the shapes?

Can you prove that you have made quarters?

5.3 Counting in halves and quarters

We are learning to count in halves and quarters.

Before we start

What numbers are missing on these number lines?

a)

0 1 2 4 6 9 10

b)

0 2 6 8 12 18 22 24

We can count on and back in halves and quarters.

Let's learn

Finlay is counting how many pizza slices he has:

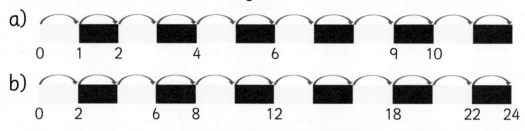

One half, two halves, three halves, four halves, FIVE HALVES.

He can count in halves using a counting stick:

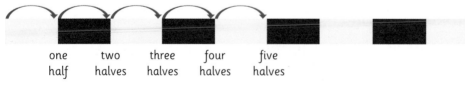

| one half | two halves | three halves | four halves | five halves |

Nuria is counting how many pizza slices she has:

One quarter, two quarters, three quarters, four quarters, five quarters, SIX QUARTERS.

She can count in quarters using a counting stick:

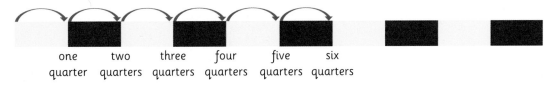

| one quarter | two quarters | three quarters | four quarters | five quarters | six quarters |

Let's practise

1) This is a whole sandwich:

one whole

Count out the sandwiches to your partner and write how many altogether (the first one has been done for you):

a) ____4 half____ sandwiches.

b) _____ sandwiches.

c) _____ sandwiches.

d) _____ sandwiches.

2) Draw counting sticks to count the following (the first one has been done for you):

a) Half pizzas:

one half   two halves   three halves   four halves   five halves

b) Quarter sandwiches:

c) Half chocolate bars:

d) Quarter oranges:

⭐ **CHALLENGE!** ....

Design number lines for your classroom wall that show how to count in:

• Ones

• Halves

• Quarters

5.4 Sharing and grouping equally

We are learning to share and group equally.

Before we start

Who has the same amount?

Finlay

Isla

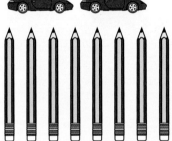

Nuria

Amman

When we share fairly everybody gets the same.

Let's learn

I have shared my sweets.

Finlay has shared his sweets but he has not **shared fairly**.

When we **share fairly**, everybody must get the **same amount**. Let's try again ...

We have an **equal** amount of sweets.

Now everybody has the same amount. Finlay has **shared** his sweets **fairly**.

Let's practise

1) Make **equal shares**:

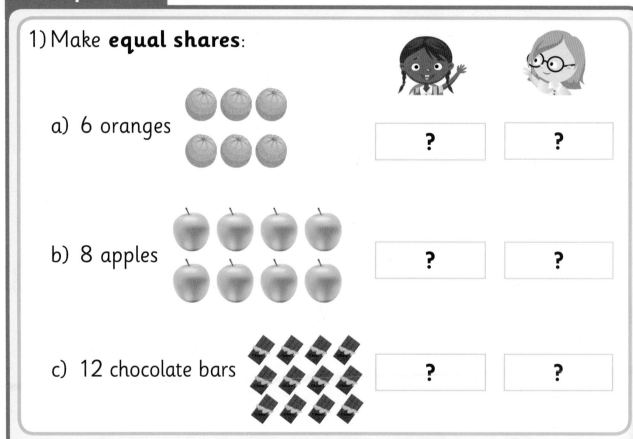

a) 6 oranges

| ? | ? |

b) 8 apples

| ? | ? |

c) 12 chocolate bars

| ? | ? |

2) Make **equal shares**:

a) 6 pizzas

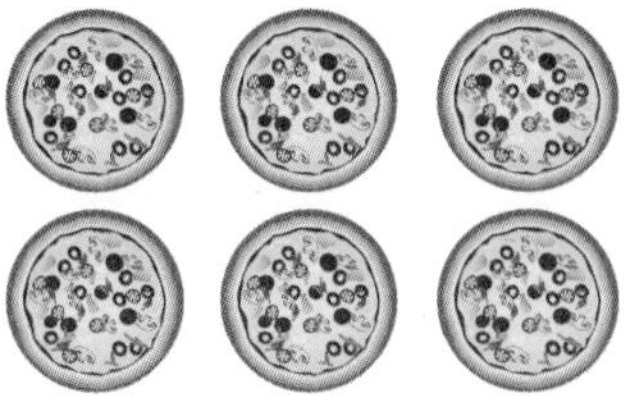

| ? | ? | ? |

6 pizzas shared between
3 makes _____ pizzas each.

b) 9 sandwiches

| ? | ? | ? |

9 sandwiches shared between
3 makes _____ sandwiches each.

c) 12 bananas

| ? | ? | ? |

12 bananas shared between
3 makes _____ bananas each.

3) a) Share 10 toy cars between 2.

| ? | ? |

b) Share 12 pencils between 3.

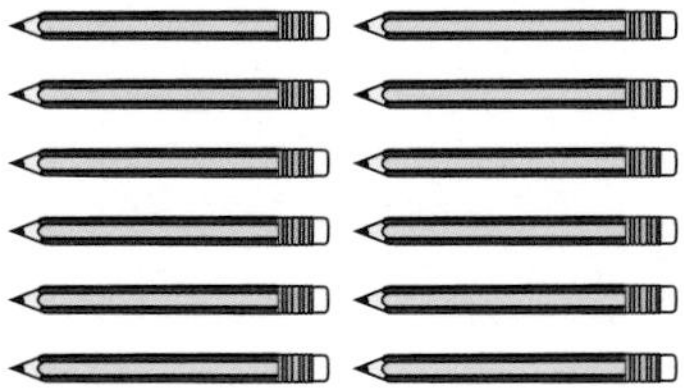

| ? | ? | ? |

c) Share 16 pound coins between 4.

| ? | ? | ? | ? |

d) Share 20 marbles between 5.

?	?	?	?	?

CHALLENGE!

17 chocolate bars can be shared equally between the 4 of us.

No they can't!

a) Who do you agree with?

b) Can you prove it in different ways?

5 Fractions, decimal fractions and percentages

5.5 Finding equivalent fractions

We are learning to find equivalent fractions.

Before we start

How much pizza do the children have?

a) Nuria

b) Amman

c) Isla

We can find fractions that are the same size.

Let's learn

Finlay and Nuria each have the same cake:

I have cut my cake into **two equal parts**. I have made **halves**.

I have cut my cake into **four equal parts**. I have made **quarters**.

Finlay eats one half of his cake.

Nuria eats two quarters.

They both eat the same amount of cake.

 is the same as

one half **is equal to** two quarters

Let's practise

1) Match the groups that have the same amount of cake.

Green group

Blue group

Red group

Orange group

Purple group

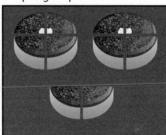

Brown group

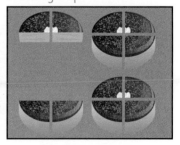

2) Match the fractions:

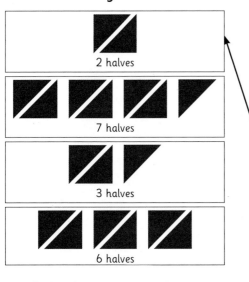

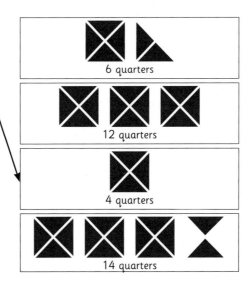

__2 halves__   is the same as   __4 quarters__

__7 halves__   is the same as   _____**?**_____

__3 halves__   is the same as   _____**?**_____

__6 halves__   is the same as   _____**?**_____

3) How many quarters can the children make? Write and draw your answers. (The first one has been done for you.)

I have 4 halves.

Finlay can make   __8 quarters__.

I have 5 halves.

Nuria can make   _____**?**_____.

I have 8 halves.

Amman can make _____**?**_____.

 I have 10 halves.

Isla can make _____?_____.

CHALLENGE!

Isla is trying to find different ways of having the same amount.

She has four halves.

 Four half pizzas is the same as two whole pizzas or eight quarter pizzas.

  =   =

Find other ways of having the same amount for each of the children.

I have six quarter oranges.

 I have three whole sandwiches.

 I have six half bars of chocolate.

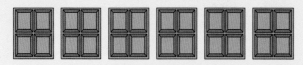

5 Fractions, decimal fractions and percentages

5.6 Share equally

We are learning to share equally (find half and quarter of an amount).

Before we start

Share equally between Finlay and Nuria.

a)

four oranges

b)

ten apples

We can find a half or quarter of an amount by sharing equally.

Let's learn

Finlay has six sweets.

Can I have **half** of your sweets please?

When we find **half** we split into two equal groups.

We have shared the sweets fairly.

We each have half of the sweets.

one half   one half

One half of six sweets is three sweets.

Isla has eight sweets.

She wants to share them fairly with her friends.

one
quarter

One quarter of eight sweets is two sweets.

The sweets have been split into four equal groups.

Each group is **one quarter** of all the sweets.

Let's practise

1) Find **one half**:

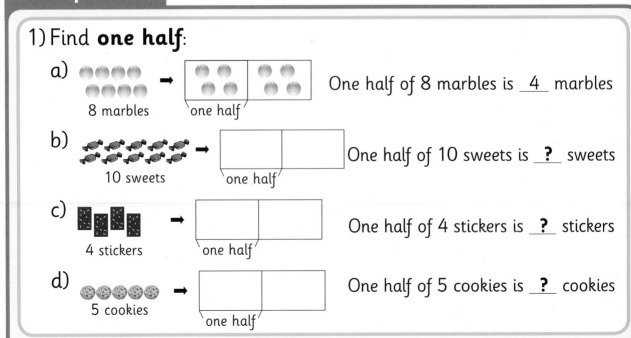

a) 8 marbles → one half   One half of 8 marbles is __4__ marbles

b) 10 sweets → one half   One half of 10 sweets is __?__ sweets

c) 4 stickers → one half   One half of 4 stickers is __?__ stickers

d) 5 cookies → one half   One half of 5 cookies is __?__ cookies

2) Find **one quarter**:

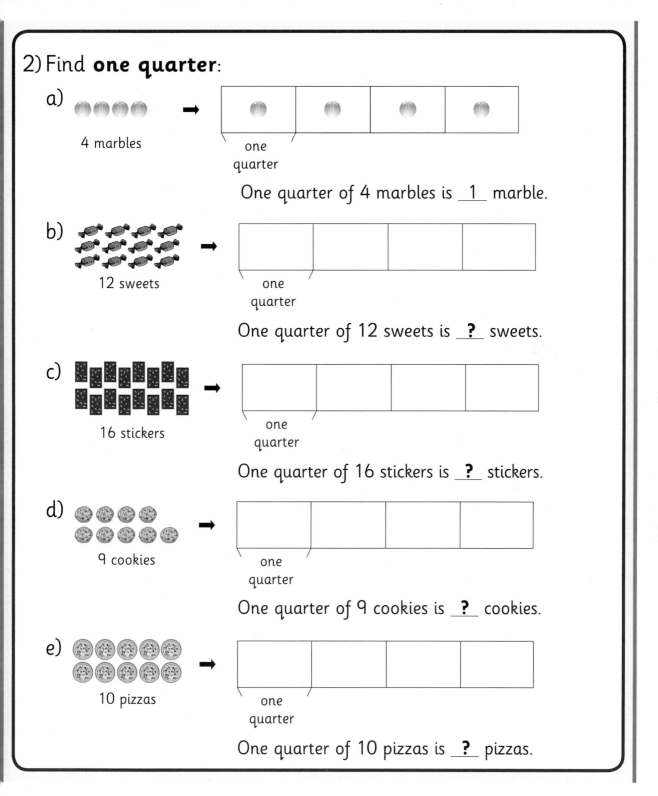

a)

4 marbles

one quarter

One quarter of 4 marbles is __1__ marble.

b)

12 sweets

one quarter

One quarter of 12 sweets is __?__ sweets.

c)

16 stickers

one quarter

One quarter of 16 stickers is __?__ stickers.

d)

9 cookies

one quarter

One quarter of 9 cookies is __?__ cookies.

e)

10 pizzas

one quarter

One quarter of 10 pizzas is __?__ pizzas.

3) How many will the children get?

I am going to get half of the sweets.

20 sweets

I am going to get one quarter of the sweets.

14 cookies

I am going to get half of the cookies.

I am going to get one quarter of the cookies.

a) Nuria will get ____ sweets.

b) Finlay will get ___ cookies.

c) Amman will get ____ sweets.

d) Isla will get ___ cookies.

★ CHALLENGE!

three quarters
Eats

one quarter
Left

1) Amman has a packet of 12 sweets. He eats three-quarters of them. How many does he eat?

2) Find three-quarters of
   a) 8 pencils
   b) 16 stickers
   c) 20 cookies

6 Money

6.1 Making amounts

We are learning to use coins to make amounts of money.

Before we start

Who has more money?
Talk to a partner.

Amman

Isla

We can put different sets of coins together to make the same amount.

We always start counting from the coin with the highest value.

Let's learn

Nuria, Amman and Isla all have 20p.

They have different coins, but the amounts all add up to 20p.

Here are Nuria's coins	Here is Amman's coin	Here are Isla's coins

1) Use these coins.

Draw the pairs of coins you need to make:

a) 7p   b) 12p   c) 6p   d) 15p

2) Use these coins. Draw your answers in your jotter.

a) Make 12p using three coins.
b) Make 20p using four coins.
c) Make 8p using four coins.
d) Make 15p using four coins.

3) Draw the coins needed to make amounts.

Try to find two different ways for each toy.

a)

b)

c)

d)

6p

15p

11p

9p

★ CHALLENGE!..

Finlay has 16p in his wallet.

a) What coins might Finlay have in his wallet?

b) If one coin is a 10p, can you find all the possible sets?

6.2 Adding amounts

We are learning to add up amounts and find the best way to pay.

Before we start

Which one adds up to 20? a)  9 + 3 + 6   b)  2 + 12 + 6   c)  11 + 5 + 3

We can add amounts to make a total. This tells us how much we have to pay.

Let's learn

Nuria buys these cakes from the school fair.

8p   6p   5p

Nuria pays 8p + 6p + 5p = **19p**.

She can pay with just four coins:

1) How much does Amman have to pay in total for each group of items?

a)

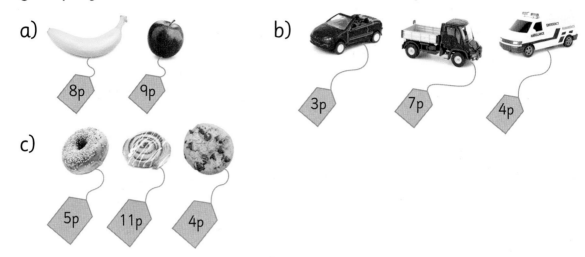

8p   9p

b)

3p   7p   4p

c)

5p   11p   4p

2) Find the best way to pay for each amount, using the least number of coins.

a) 12p (use two coins).

b) 8p (use three coins).

c) 17p (use three coins).

d) 13p (use three coins).

3) Add up the totals. Draw the least number of coins to pay for each amount in your jotter.
   Remember to start with the coin with the highest value.

a) 13p + 7p = [?]

Coins: [?]

b) 4p + 1p + 2p = [?]

Coins: [?]

c) 9p + 5p = [?]

Coins: [?]

d) 9p + 9p = [?]

Coins: [?]

⭐ CHALLENGE! ⋯⋯⋯⋯⋯⋯⋯⋯⋯⋯⋯⋯⋯⋯⋯⋯⋯

The children are in a sweet shop.
Find the total that each child pays.
Then find the best way to pay, using the least number of coins.

12p	9p	4p	2p	11p

a) Nuria buys

b) Isla buys

c) Amman buys

d) Finlay buys

6 Money

6.3 Calculating change

We are learning to give change up to 20p.

Before we start

Which of these numbers, take away 4, gives the answer 12?

15 16 17 18

When we don't have the right coins to pay, we can get change.

Let's learn

Finlay buys a tractor.

It costs 14p.

14p

Finlay has only a 20p coin.

The shopkeeper gives Finlay change. He counts on from 14p.

Finlay gets 6p change.

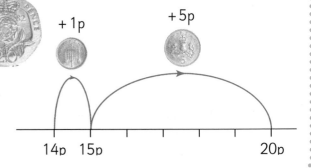

+1p   +5p

14p 15p   20p

Let's practise

1) The children are buying ice lollies. They each have a 10p coin. Draw the change they get.

a)
| 6p | Nuria's lolly

b)
| 7p | Isla's lolly

c)
| 3p | Amman's lolly

2) Isla has these coins:

a) Talk to a partner. For each fruit, decide if Isla can give the exact money, or if she will need change.
Write your answer in your jotter.

b) If Isla needs change, draw the coins she will get.

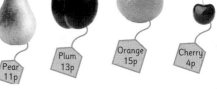

 Pear 11p   Plum 13p   Orange 15p   Cherry 4p

3) Finlay has a 20p coin.

Draw the change he will get if he buys things that cost these amounts:

a) 15p

b) 17p

c) 12p

d) 11p

CHALLENGE!

Amman has one 20p coin. He buys all these toys.

 5p

 3p

 3p

a) How much does Amman need to pay?

b) How much money does he have left?

7.1 Days of the week

We are learning to place tasks into a daily timetable and record dates accurately.

Before we start

1) Place the following days of the week in order: Tuesday, Saturday, Monday and Wednesday.

2) Copy and complete the following list of days:
   Sunday, ____?____, Tuesday, ____?____,
   Thursday, ____?____, Saturday.

3) What is the day *before* and *after*:
   a) Tuesday   b) Saturday

A **timetable** shows a list of days and times to tell us when something happens.

Let's learn

Timetables are lists of days and times that let us know when an event is going to take place.

A school timetable will show us what day you will have certain lessons and at what time.

	MONDAY	TUESDAY	WEDNESDAY	THURSDAY	FRIDAY
9.00–10.30	Literacy	Numeracy	Literacy	Numeracy	Literacy
10.30–10.45	←		BREAK		→
10.45–12.15	Numeracy	Literacy	Numeracy	Literacy	Numeracy
12.15–1.00	←		LUNCH		→
1.00–2.00	P.E.	HWB	Assembly	Science	RME
2.00–3.00	Art	RME	IDL	P.E.	HWB

Times of the lessons each day.   Different lessons during the day.

	MONDAY	TUESDAY	WEDNESDAY	THURSDAY	FRIDAY
9.00–10.30	Literacy	Numeracy	Literacy	Numeracy	Literacy
10.30–10.45	BREAK	BREAK	BREAK	BREAK	BREAK
10.45–12.15	Numeracy	Literacy	Numeracy	Literacy	Numeracy
12.15–1.00	LUNCH	LUNCH	LUNCH	LUNCH	LUNCH
1.00–2.00	Art	P.E.	Free Play	P.E.	RME
2.00–3.00	RME	Assembly	Science	Outdoor Play	Science

1) What day and time is Art?

2) How long do the class have P.E. every week?

3) Isla loves Science! What days will she like the best?

4) If Assembly swaps time with Outdoor Play, what day and time will it now be on?

CHALLENGE! ....

Work with a partner to copy and complete the timetable to record your school day.

Remember to include:

Blank timetables

✓ Times

✓ Break and Lunch

✓ All subjects

Extension:

Now complete your weekly timetable in the same way.

7.2 Measuring time

> We are learning to estimate and time the duration of an activity.

Before we start

Sort these activities starting with the ones that take the **least** time to the ones that will take the **most** time:

Brushing teeth   Walking to school   Eat breakfast   Jump on the spot 10 times

> You can measure time using prior knowledge.

Let's learn

If you **estimate** how long something will take, you will use your prior knowledge about the duration of an activity.

If you **time** something, you will use a device like a clock, watch or stopwatch to measure how long the activity will take.

If you walk to school every day, then you might know how long it takes you by looking at the **clock** or your **watch** before you leave the house and then when you get to school.

1) Match the activity with the best unit of time to measure it:

Hop 10 times	Grow a sunflower	Get dressed	Run a marathon

Hours	Minutes	Days	Seconds

2) Match the timing device to its time unit:

Hours	Days	Seconds	Minutes

CHALLENGE!

Work with a partner to measure time.
Use a stopwatch to copy and complete the following table:

	Estimated time	Actual time
Write your name ten times		
Throw and catch a ball ten times		
Write the numbers 1 to 10 five times		
Walk round the classroom ten times		

7.3 Read o'clock and half past

We are learning to read o'clock and half past.

Before we start

1) Copy the clock face and write in the missing numbers.

2) Label the minute hand and the hour hand.

I am learning to read the time at half past the hour.

Let's learn

If the hour hand is pointing directly at a number and the minute hand is pointing straight up, then it is a whole hour time.

If the hour hand is halfway between two numbers and the minute hand is pointing straight down, then it is half past an hour.

Let's practise

1) On these clocks, colour the **hour hand** blue and colour the **minute hand** red.

2) Match the clocks to the times.

2 o'clock

Half past 12

Half past 4

11 o'clock

6 o'clock

Half past 7

3) Amman must be home by 4 o'clock. This is Amman's watch.

   Is he late? Write yes or no in your jotter.

4) a) This is 7 o'clock. Draw arrows on the clock to show which way the hands are moving.

   b) Draw where the hands will be at half past 7.

CHALLENGE! ....

Read Isla and Nuria's clues carefully and then draw the time.

Half an hour has gone by since 9 o'clock.

There is half an hour to go until 10 o'clock.

Make up two clues to give to a friend. See if they can guess the time you are thinking of.

7.4 Change o'clock and half past from analogue to digital

> We are learning to convert o'clock and half past from analogue to digital and vice versa.

Before we start

1) What time is shown on the clocks?
   Write your answers in your jotter.

a)

b)

c)

d)

e)

f)

> Time can be displayed in analogue or digital.

Let's learn

An **analogue clock** has hands that move round to point to the time.

A **digital clock** has no hands.

The numbers change to show the time.

Blank clock faces

Let's practise

1) Match the times.

Resource 1A_7.4_Let's practise_Q1

| 3:00 | 9:00 | 7:00 | 1:00 |

2) Match the times.

Resource 1A_7.4_Let's practise_Q2

| 3:30 | 2:30 | 9:30 | 10:30 |

3) Write the times in your jotter.

a) 11 : 30   b) 8 : 00   c) 3 : 30   d) 02 : 00

Half past ?   ? o'clock   Half past ?   ? o'clock

CHALLENGE!

Fill in the times and missing clock hands.

a)

? : ?   Half past ___?___

Resource 1A_7.4_Challenge

b)

06 : 00   ___?___ o'clock

8 Measurement

8.1 Comparing and ordering lengths and heights

We are learning to compare and order lengths and heights.

Before we start

a) Which tower is taller?

i)   ii)

b) Which pencil is shorter?

i)   ii)

We can compare lengths and heights using **longer**, **taller** and **shorter**.

Let's learn

Amman and his mum want to ride on the roller coaster.

You must be this tall to ride.

Mum   Amman   boy

Mum is **taller** than the boy. She **can** ride on the roller coaster.

Amman is **shorter** than the boy. He **cannot** ride on the roller coaster.

The children are comparing the length of their hair:

My hair is **shorter** than Isla's.

My hair is **longer** than Finlay's.

I have the **longest** hair.

Finlay has the **shortest** hair.

Nuria has the **longest** hair.

Let's practise

1) Write the missing word in your jotter.

bike   car

blue
red

a) The __?__ is longer.

b) The __?__ bar is shorter.

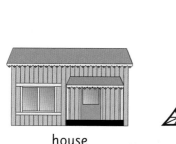

house   tower

boy   man

c) The __?__ is taller.

d) The __?__ is shorter.

2) Write these animals in order from shortest to tallest in your jotter:

horse   cow   giraffe   cat   elephant

3) Draw five or six towers in your jotter, using a different colour for each.

Make sure they are all different heights like these:

Write six sentences comparing the heights of the towers in your jotter. For example:

*The blue tower is **smaller than** the red tower.*

Get into a group of four or five with your classmates.

Order yourselves in a line from shortest to longest.

In your jotter, draw a picture of the line of classmates in order and write their names.

Write three or four sentences that compare the heights.

8 Measurement

8.2 Comparing and ordering mass

> We are learning to compare and order mass.

Before we start

Which is heavier?

 or or or

Water bottle   Tennis ball   Balloon   £1 Coin

> We can use words like heavier and lighter to compare mass.

Let's learn

Nuria and her mum sit on the see-saw.

> I have gone up and you have gone down. I am lighter than you.

> Which is heavier?

> The teddy is **bigger**. I think it will be **heavier**.

The children use a pan balance to find out.

The balance shows us that the book is heavier.

It doesn't matter that the teddy is bigger.

Let's practise

1) Which is heavier? Write the answers in your jotter.

a)
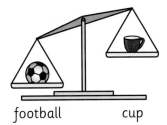

football   cup

b)
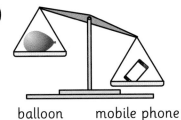

balloon   mobile phone

c)
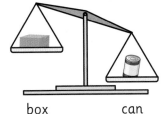

box   can

d)
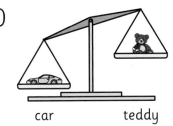

car   teddy

2) Complete each sentence in your jotter, saying if the object is heavier or lighter.

a)

The apple is ___?___.

b)
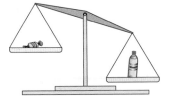

The water is ___?___.

c)

The tablet is ___?___.

d)

The pencil case is ___?___.

3) Draw the scales in your jotter. What colour are the missing boxes.

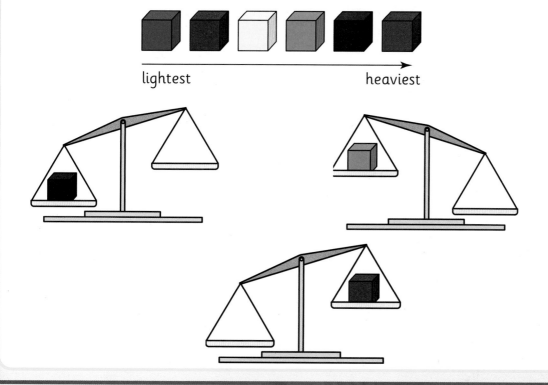

lightest → heaviest

The children compare the weights of these objects using the pan balance.

 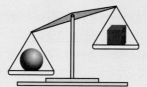 

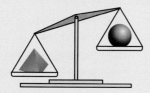 

Write or draw the shapes in order from lightest to heaviest.

8.3 Comparing and ordering area

We are learning to compare and order area.

Before we start

a) The football pitch is _____?_____ than the tennis court.

Football pitch

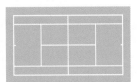

Tennis court

b) The garden is _____?_____ than the park.

garden

park

We can use words like larger and smaller to compare area.

Let's learn

Isla and Amman are each painting a wall.

I have more space to fill on my wall. I will need more paint than you.

Amman has a larger area to paint.

The children are helping to cut the grass in their gardens.

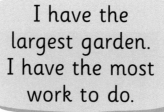
Can we order the size of our gardens from smallest to largest?

I have the largest garden. I have the most work to do.

I have the smallest garden. I have the least work to do.

We can order the children from smallest to largest garden:

smallest
garden

largest
garden

Let's practise

1) In your jotter, write the names of the six rooms in order of size from smallest to largest.

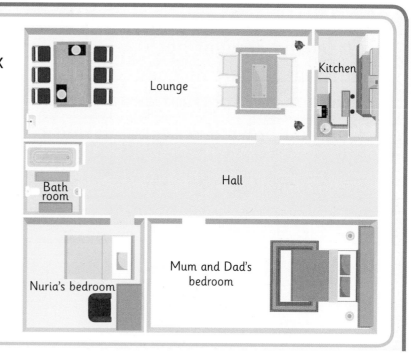

Lounge

Kitchen

Bath room

Hall

Nuria's bedroom

Mum and Dad's bedroom

2) Match the tablecloths to the correct tables:

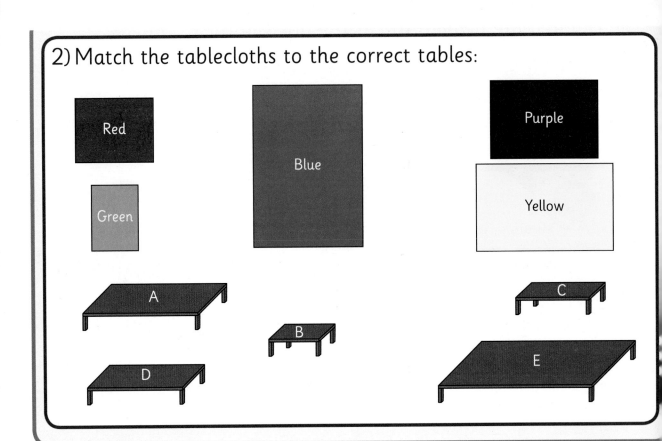

CHALLENGE!

These patterns have been made using red, yellow and blue paint:

a) Order the amount of different colours of paint used for each pattern, from least to most.

Can you prove it by copying and cutting out the shapes?

b) Make your own pattern using different-sized coloured paper. Challenge your partner to order them from smallest to largest.

8.4 Comparing and ordering capacity

We are learning to compare and order capacity.

Before we start

We can use words like **larger**, **smaller**, **more** and **less** to compare capacity.

Which holds more?

a)

or

kettle   mug   glass   container

b)

or

Let's learn

Capacity is the amount that a container can hold.

The box can hold more cereal than the bowl.

The box has a larger capacity than the bowl.

The glass can hold less water than the jug.

The glass has a smaller capacity than the jug.

The hot air balloon holds more air than the balloon.

The hot air balloon has a larger capacity than the balloon.

1) Which has the larger capacity?

a)

sink   bath

b)

box   bag

c)

mug   kettle

d)

can   saucepan

2) Copy and complete:

a)

The can holds _____?_____ juice than the bottle.

b)

The van has _____?_____ space than the car.

c)

The bin bag holds _____?_____ rubbish than the wheelie bin.

d)

The pool holds _____?_____ water than the bath.

3) Order from the smallest to largest capacity:

car boot

bin bag

rubbish truck

cereal box

wheelie bin

shopping bag

snack tub

⭐ **CHALLENGE!** ....

Sweets come in two sizes of boxes like these (cubes):

Amman can choose one green box or four red boxes of sweets.

a) Which should he choose if he wants more sweets?

b) Can you prove it?

8.5 Estimating and measuring length (non-standard units)

We are learning to estimate and measure length using non-standard units.

Before we start

a) Which is shorter?

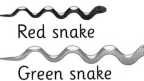

Red snake

Green snake

b) Who is taller?

Dad   Nuria

We can use any object to measure length or height.

Let's learn

Nuria uses cubes to measure the length of her pencil:

The pencil is about four cubes long.

That's not right. You need to line them up like this...

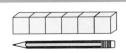

Nuria's pencil is six cubes long.

Finlay is estimating Isla's height:

I think you are as tall as five building blocks.

Let's check!

Finlay uses the building blocks to check:

You are as tall as four building blocks.

1) Estimate the length of the tables for Isla:

   a) __4__ pencils long

   b) __?__ pencils long

   c) __?__ pencils long

   d) __?__ pencils long

   e) __?__ pencils long

   I think the brown table will be four pencils long.

2) Use cubes to measure the length of each of these bars:

   a) ____?____ cubes

   b) ____?____ cubes

   c) ____?____ cubes

   d) ____?____ cubes

CHALLENGE!

Choose four or five objects from the environment around you. Use cubes to estimate and measure their length or height. Record your measurements in a table like this:

Object	Estimate	Length or height

8.6 Estimating and measuring mass (non-standard units)

We are learning to estimate and measure mass using non-standard units.

Which is heavier?

a)

candle   mobile phone

b)

magazine   book

We can use smaller objects to measure the weight of larger objects.

Let's learn

Amman uses beanbags to measure the weight of his pencil case:

 The pencil case balances with the beanbags.

The pencil case weighs the **same as** three beanbags.

The children use beanbags to measure the weight of the lunchbox:

 The lunchbox weighs more than four beanbags.

The lunchbox weighs less than six beanbags.

We can estimate that the lunchbox weighs about the same as five beanbags.

Let's practise

1) Write a sentence to describe the weight of each item.

a)

The shoe weighs _____?_____ beanbags.

b)

The mobile weighs _____?_____ beanbags.

c)

The tablet weighs _____?_____ beanbags.

2) Estimate the weight of each item.

a)

The carrot weighs _____?_____.

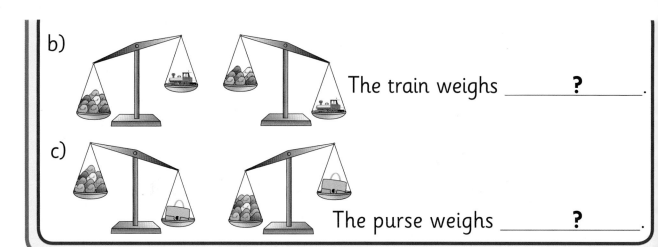

b)

The train weighs _____?_____.

c)

The purse weighs _____?_____.

★ CHALLENGE! ....

You will need:

- A pan balance:

- Four or five items to weigh:

- Items to measure with (for example, cubes or dominoes):

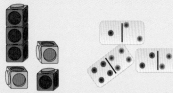

a) Estimate the weight of each item.

b) Weigh each item.

Record your measurements in a table like this:

Object	Estimate	Weight

8.7 Estimating and measuring area (non-standard units)

We are learning to estimate and measure area using non-standard units.

Before we start

Order these areas from largest to smallest:

- Gym hall
- Playground
- Cover of this book
- Desk
- Classroom

We can estimate and measure area by covering or filling with smaller objects.

Let's learn

Nuria and Finlay are setting out the food for a party:

How many trays can fit on the table?

The table has enough space to fit six trays.

I think five trays will fit. Let's put them on to see.

Let's practise

1) Estimate how many will fit:

a) Slices of toast on the tray

b) Books on the table

c) Tins of beans in the box

d) Pairs of shoes on the newspaper

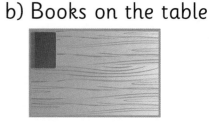

e) Socks over the radiator

2) Estimate and measure the following. How many fit?

a) Books on a desk

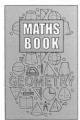

b) Counters on a sheet of paper

c) Cubes on a plate

d) Shapes on a book cover (try different shapes)

e) Hands on a cupboard door

CHALLENGE!

You will need:

- Tape or chalk
- To work in a group

1) Mark out two or three different large shapes on the floor.

2) Estimate how many children will be able to lie in each.

3) Check.

Can you mark out shapes that will fit exactly:

- Two children
- Four children
- Eight children

8.8 Estimating and measuring capacity (non-standard units)

We are learning to estimate and measure capacity.

Before we start

1) Order these containers from smallest to largest capacity:

a) b) c) d) e)

We can use any smaller containers to measure the capacity of larger containers.

Let's learn

Nuria is finding out how many cups it will take to fill up the jug:

It takes five cups of water to fill the jug.

Amman uses glasses of water to measure the capacity of the pot:

The pot can hold four glasses of water.

You must fill each glass, like this.

   ×  ✓

The pot holds three FULL glasses of water.

1) Estimate how many:

a)

b)

c)

d)

2) You will need:
- A jug
- Some different containers

Resource 1A_8.8_Let's practise_Q2

How many of each container will it take to fill the jug with water? (Try estimating first.)

Container	Estimate	Actual
yoghurt pot		
cup		
jar		
bottle		

You will need:

- Three or four different-sized containers
- A plastic cup.

Fill the containers to different levels.

a) Challenge your partner to estimate how many cups of water will be needed to fill each container.

b) Fill the containers to check.

9.1 Ways numbers are used in real life

We are learning the ways that numbers are used in real life.

Before we start

With a partner, tell each other where you see and use numbers.

There are objects with numbers on them all around us.

Let's learn

At the shops you will see numbers on the price tags and at the end of each aisle. We use this information to make choices.

Let's practise

1) Put these speed limits in order, from smallest to biggest.

a) **15**

b) **20**

c) **5**

d) **10**

e) **70**

f) **50**

g) **60**

h) **40**

2) Match the picture to the object.

a) Bus

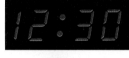

b) Calculator

c) Phone

d) Clock

e) TV remote control

f) Oven

g) Recipe book

h) Door number

i) Library book

CHALLENGE!

Try to find 10 different number displays in the room you are in.
Challenge a partner to find more!

10 Patterns and relationships

10.1 Continuing patterns

We are learning to recognise and continue patterns.

Before we start

Look at the pattern. Can you describe it to your partner?

Patterns of number and shape follow a rule.

Let's learn

Patterns can be arranged in many different ways including shape, colour and number.

          Shape pattern

          Colour pattern

2, 4, 6, 8   Number pattern

1) Copy and complete the following patterns in your jotter:

a)  ?

b)   ?

c) 1, 2, 1, 2, ?

2) Copy and complete the following patterns in your jotter:

a) ? ?

b)   ? ?

c) 1, 1, 2, 1, 1, ? ?

CHALLENGE!

Using the different socks in the picture, how many different patterns and sequences can you create?

10.2 Recognise and continue number sequences

We are learning to recognise and continue number sequences.

Before we start

Copy and complete the following patterns in your jotter:

a)

b)

We need to look for a pattern to find the missing number in a sequence.

Let's learn

In a sequence, things follow on one after the other.

This is a sequence of odd numbers:

1, 3, 5, 7, ...

If there are missing numbers in a sequence you need to look for a pattern.

The numbers might be getting bigger	**5, 10, 15...**
The numbers might be getting smaller	**10, 8, 6...**
You might be adding on the same number each time	**1, 4, 7, 10... (+3)**
You might be taking away the same number each time	**9, 7, 5, 3... (−2)**

1) Copy and complete the number sequences in your jotter:

   a) 0, 2, 4, 6, [?] [?]   b) 2, 5, 8, 11, [?] [?]

   c) 23, 21, 19, 17, [?] [?]

2) Copy the number sequences. Fill in the missing numbers in your jotter:

   a) 1, [?] , 5, [?] , 9   b) 20, [?] , 10, [?] , 0

   c) 10, 20, [?] , [?] , 50

⭐ CHALLENGE! ....

100 square

Work with a partner using a hundred square to find as many number sequences as you can.

You will need to be able to explain the patterns that you have found.

1	2	3	4	5	6	7	8	9	10
11	12	13	14	15	16	17	18	19	20
21	22	23	24	25	26	27	28	29	30
31	32	33	34	35	36	37	38	39	40
41	42	43	44	45	46	47	48	49	50
51	52	53	54	55	56	57	58	59	60
61	62	63	64	65	66	67	68	69	70
71	72	73	74	75	76	77	78	79	80
81	82	83	84	85	86	87	88	89	90
91	92	93	94	95	96	97	98	99	100

Colour in each number pattern.

11 Expressions and Equations

11.1 Equations

We are learning to complete a number sentence.

Before we start

Say if the following are **true** or **false**:
a) 1 + 9 = 11   b) 14 = 7 + 7
c) 3 + 3 = 2 × 3   d) 3 + 7 = 9 + 2

Equations are number sentences.

Let's learn

Equations follow the same **pattern** as number sentences.

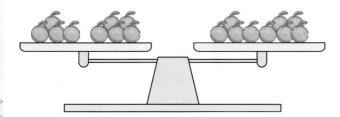

4 oranges + 5 oranges = 9 oranges

Both sides are **equal**.

Sometimes there is a missing number that you will have to find the value of.

5 + ☐ = 15   5 + **10** = 15

1) Copy and complete the following number sentences in your jotter:

a) $8 + \boxed{?} = 16$

b) $15 - \boxed{?} = 5$

c) $\boxed{?} + 11 = 20$

d) $3 + 9 = \boxed{?}$

e) $6 + \boxed{?} = 12$

f) $\boxed{?} = 8 + 2$

g) $20 = 4 + \boxed{?}$

h) $\boxed{?} - 6 = 14$

i) $13 = 7 + \boxed{?}$

2) If △ = 5 and ● = 2, copy and complete the following equations in your jotter by finding the missing numbers:

a) $\boxed{?} + △ = 10$

b) $\boxed{?} + ● = 10$

c) $△ + \boxed{?} = 15$

d) $\boxed{?} - ● = 6$

3) Copy and complete the following equations in your jotter:

a) $20 \boxed{?} 10 = 10$

b) $5 \boxed{?} 8 = 13$

c) $9 \boxed{?} 5 = 4$

d) $6 \boxed{?} 6 = 12$

★ CHALLENGE!

Find the value of the shapes shown by solving the equations in your jotter.

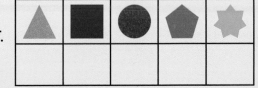

1) $● + 2 = 8$

2) $2 \times 5 = ✦$

3) $4 + 4 = ⬠$

4) $14 - 10 = ■$

5) $△ + 10 = 12$

12 2D shapes and 3D objects

12.1 2D shapes

> We are learning to name and describe 2D shapes.

Before we start

Which is the odd one out? Explain why.

a) 　　b) 　　c) 　　d)

> Different shapes have different properties.

Let's learn

Square

Triangle

Rectangle

Circle

1) Isla is describing shapes to Nuria. Help Nuria to name each shape.
   a) It has three sides. It is a ...
   b) It has four sides that are all the same length. It is a ...
   c) It is round. It is a ...
   d) It has four sides, but they are not all the same length. The opposite sides *are* the same length. It is a ...

2) Which is the odd one out?
   Write a sentence in your jotter to explain why.
   a)
   b)
   c)

3) Some of these shapes are triangles. Find the shapes that are not triangles. Explain your thinking to a partner.
   a)
   b)
   c)
   d)
   e)

CHALLENGE!

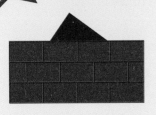

There is a shape hiding behind the wall. Nuria says it could be a square. Amman thinks it must be a triangle. Who is right? Talk to a partner.

12.2 Sorting 2D shapes

We are learning to sort 2D shapes by their properties.

Before we start

 Which of these shapes will fit in the hole?

a) 　　　b) 　　　c)

We can sort shapes by looking at how many **sides** and how many **corners** they have.

Let's learn

All shapes have sides.

Some shapes have straight sides.

Some shapes have curved sides.

Most shapes have corners. Can you think of a shape with no corners?

1) Look at these shapes. Sort them by the number of corners they have.

   a) Colour the shapes with no corners purple.

   b) Colour the shapes with three corners orange.

   c) Colour the shapes with four corners green.

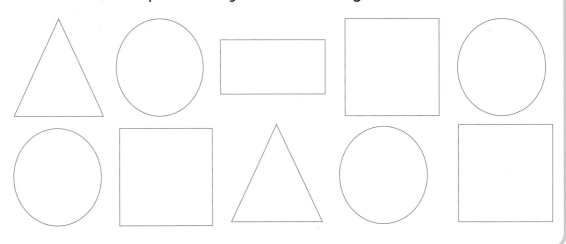

2) These shapes have been sorted. Discuss what the sorting rule might be with a partner.

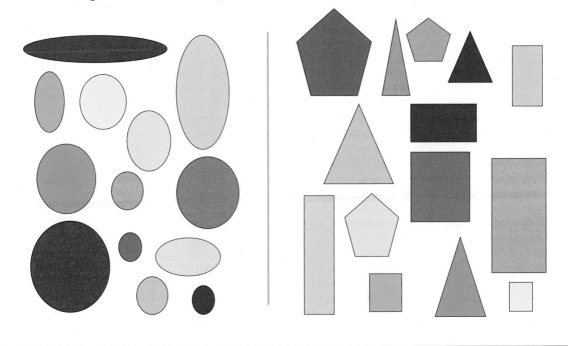

3) Look at these shapes.

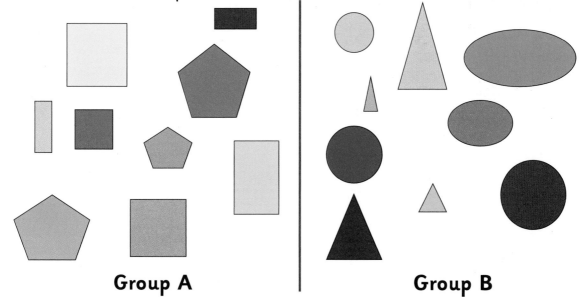

Group A   **Group B**

a) What is the sorting rule? Discuss this with a partner.

b) Look at group A. Can you sort the shapes further?

⭐ CHALLENGE! ....

Sort these shapes using this sorting rule.

Copy and complete the table below.

Four sides	Four corners

a) Did you have any problems doing this? Why did the sorting rule not work? Discuss this with a partner.

b) Can you think of a better sorting rule that you could use to sort these shapes? Write it in your jotter and sort the shapes again.

12.3 Sorting 3D objects

We are learning to sort 3D objects
by their properties.

Before we start

Which is the odd one out?
Explain your thinking.

We can sort 3D objects by looking at how
many **faces, edges** and **corners** they have.

Let's learn

All 3D objects have faces.

Some 3D objects have flat faces.

Some have curved faces. We call these
curved surfaces.

We can count the faces or surfaces, edges and the corners to
help name and sort 3D objects.

1) Look at these 3D objects.

i) ii) iii) iv)

Resource 1A_
12.3_Let's
practise_Q1

cylinder   cube   sphere   cuboid

Copy the tables below. For each sorting rule, write the name or names of the objects in the correct box.

a)

Has six faces	Has fewer than six faces

b)

Has curved surfaces	Has no curved surfaces

c)

Has square faces	No faces are square

d)

Has curved surfaces and flat faces	Has flat faces

2) Here are some 3D objects.

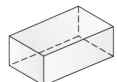

Resource 1A_
12.3_Let's
practise_Q2

Copy and complete the table below.
Sort them by the number of edges.

Five edges or more	Less than five edges

3) Finlay has sorted some 3D objects by whether or not they have corners:

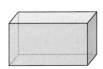

Group A   **Group B**

Which group should Finlay put the cone in?

Explain your thinking.

★ CHALLENGE! ....

Work with a partner.

Find some 3D objects in your classroom.

Each of you sort some objects by different sorting rules, but don't tell your partner your rule.

Show each other your sorted objects.
Can you work out your partner's sorting rule?

12.4 3D objects

> We are learning to name 3D objects and describe their faces, corners and edges.

Before we start

a)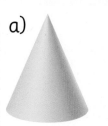

b)

c)

d)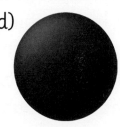

Which of these 3D objects will roll? Talk to a partner.

> We can describe a 3D shape using its properties.

Let's learn

Here are some 3D objects. The dotted lines show edges you can only see if you turn the object around.

cube

cylinder

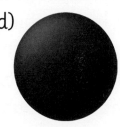

cuboid

cone

sphere

square-based pyramid

Let's practise

Use actual 3D objects to help you answer these questions.

1) For each 3D object, write down in your jotter the number of faces or surfaces.

a) 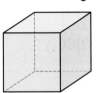   b)   c) 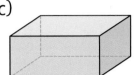   d)

2) Finlay is describing the faces of some 3D objects. Name each object from its faces or surfaces.

a) It has one curved surface. It is a …

b) It has two round, flat faces and one curved surface. It is a …

c) It has six flat faces. All the faces are square. It is a …

d) It has one round flat face, and one curved surface. It is a …

3) a) Isla's object has one corner and one edge. What is it?

b) Nuria's object has five corners and eight edges. What is it?

4) Which is the odd one out? Write a sentence to explain your thinking.

a)   b)   c)   d)

Amman wants to paint every face of his cube, square-based pyramid and cylinder. If he wants to paint each face a different colour, how many colours will he need?

13 Angles, Symmetry and Transformation

13.1 Using positional language

We are learning to describe the position of an object.

Before we start

Which is Finlay's **left** hand?   a)   b)

We use position words to say where things are in real life or on a map or a plan.

Let's learn

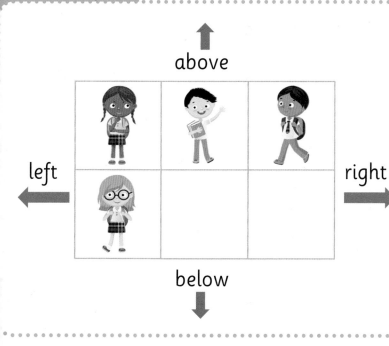

Nuria is to the **left** of Finlay.

Amman is to the **right** of Finlay.

Finlay is **between** Nuria and Amman.

Isla is **below** Nuria.

Nuria is **above** Isla.

1) Jason the Janitor has three shelves in his workshop. The map shows what he keeps on each shelf.

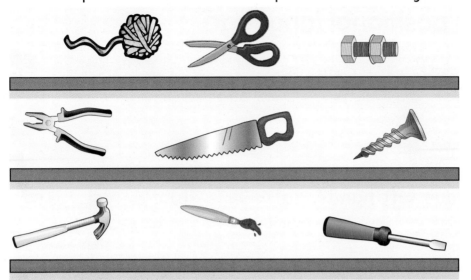

Look at the map of Jason's shelves. Draw in your jotter what is:

a) on the right of the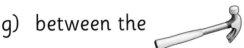

b) on the left of the

c) on the right of the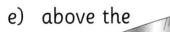

d) below the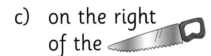

e) above the

f) below the

g) between the and the

2) On a copy of the grid shown below draw each shape in the correct box. Colour the shapes.

a) The ▨ is one square to the right of the ▪.

b) The ▨ is two squares above the ▬.

c) The ▲ is two squares to the left of the ◯.

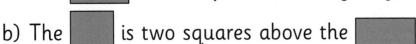
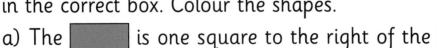
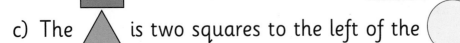

d) The 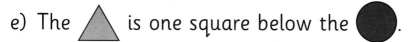 is between the ◯ and the △.

e) The △ is one square below the ⬤.

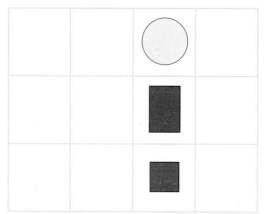

3) Work with a partner. Take turns to point to a shape on the grid in question 2. The other person must use position words to say exactly where that shape is.

CHALLENGE!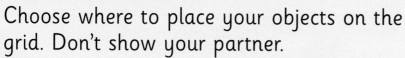

Work with a partner. Take turns to be the describer.

Use the photocopiable sheet or square paper and the same set of objects each.

Choose where to place your objects on the grid. Don't show your partner.

Keeping your grid hidden, describe where you have put the objects on your grid

Your partner must try to put their objects in the same places on their grid.

Compare your grids to see how well you did.

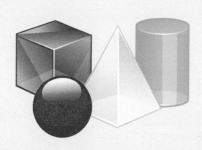

Resource 1A_13.1_Challenge

13 Angles, Symmetry and Transformation

13.2 Using directions

We are learning to follow and give directions.

Before we start

Dog   Cat

Which animal is on the left?

Forward and **backward**, and **left** and **right** are all direction words.

Let's learn

When we give directions, we need to think about where the other person is facing.

Their left and right may not be the same as ours.

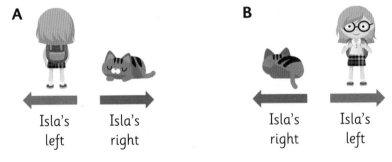

A

Isla's left   Isla's right

B

Isla's right   Isla's left

The cat is on **Isla's right** in both diagrams. But in diagram B, it is on **your left**. This is because Isla is facing the opposite direction to you.

1) This is a plan of Jollytown.

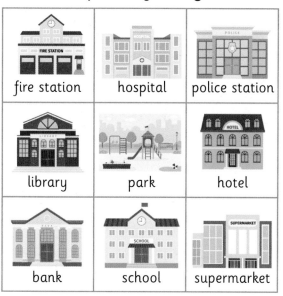

fire station	hospital	police station
library	park	hotel
bank	school	supermarket

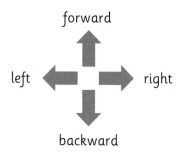

forward

left ← → right

backward

Resource 1A_13.2_Let's practise_Q1

a) Use a counter to help you follow the directions. Write in your jotter where you will end up.

i) Start at [school]. Go forward one square – left one square. You are at the [**?**].

ii) Start at [fire station]. Go backward two – right two. You are at the [**?**].

iii) Start at [hotel]. Go forward one – left one – backward one. You are at the [**?**].

iv) Start at [park]. Go forward one – right one – backward two – left one. You are at the [**?**].

b) Write down directions to get

i) From to   ii) From to

iii) From  to then to

2)

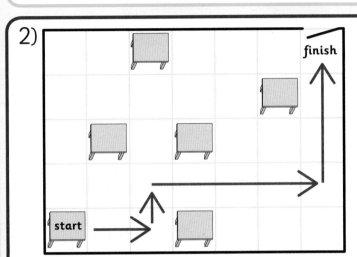

This is the path Nuria took from her table to the door.

Resource 1A_13.2_Let's practise_Q2

a) Use a counter to help you follow the directions. Think carefully about left and right. You may want to rotate the book to help.

Walk forward two squares – turn left – walk forward one square – turn right – walk forward four squares – turn left – walk forward three squares.

b) Draw a different path that Nuria could take.

Write directions for a partner to follow.

⭐ CHALLENGE! ....

Continue the directions to guide Amman through the maze.

Forward two – turn left – ...

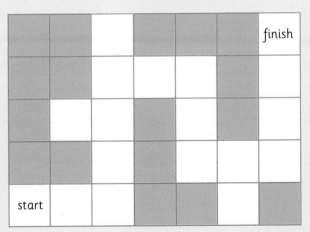

Resource 1A_13.2_Challenge

13.3 Symmetry

We are learning to create symmetrical pictures.

Before we start

a)   b)   c)   d)

Which animal is not symmetrical? Talk to a partner.

In a symmetrical picture, both sides are the same.

Let's learn

This ladybird is symmetrical.

✓   X

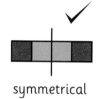

symmetrical   not symmetrical

✓   X

The pattern matches when the picture is folded along the middle.

Let's practise

1) Work with a partner.

The aim of this question is to create a symmetrical picture like this:

Use flat shapes and a piece of paper with a straight line down the middle. Or go outside and use leaves and sticks and whatever else you can find.

Player 1 places a shape one side of the line.

Player 2 places a shape on the other side of the line so that the picture stays symmetrical.

Take it in turns to start.

2) Copy and complete these patterns so they are symmetrical.

a)

b)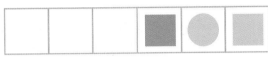

c)

d)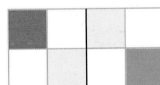

Make your own symmetrical pattern in your jotter.

3) Copy and complete to make each design symmetrical.

a)

b)

c)

d)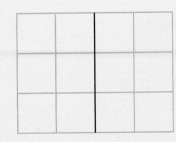

a) Can you colour the squares to make a symmetrical picture?

b) How many different symmetrical pictures can you make using the same grid and just two colours?

Blank symmetry grid

14.1 Understanding data and information

We are learning to collect data.

Before we start

How would you sort these objects?

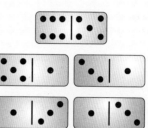

We can collect data in different ways.

Let's learn

Information can be collected in lots of different ways.

You can ask questions and record the answers in a survey or chart.

Resource 1A_ 14.1_Let's learn

Cereal	Tally
Banana Bran	卌 |||
Chock Rings	卌 ||||
Ricebixs	卌 ||
Fruity Diamonds	卌	
Berry Clusters	卌 ||

You can use tally marks to record information

1) Here are the results of a survey about favourite fruit in the tuck shop.

Copy and complete the table.

Fruit	Tally	Total
🍌		
🫐		
🍎		
🍊		
🍐		

The most popular fruit is: _____.

2) Find out if hot or cold lunch is the most popular in your class.

Type of lunch	Tally	Total
🍲		
🍱		

CHALLENGE!

With a partner, collect information about the most popular ice-cream flavour.

Create a survey and record the answers.

14.2 Understanding data and information

We are learning to understand data.

Before we start

a) How many red sweets are there?

b) How many yellow sweets are there?

c) How many green sweets are there?

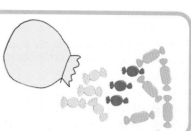

I am learning to identify values shown on a graph.

Let's learn

Graphs help us answer questions about information.
Take care to place objects onto the graph in the right place.

Let's practise

1) Here are the sweets arranged in a graph by colour.

   Which colour has the greatest number of sweets?

Colour of sweets

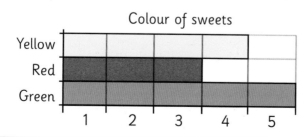

2) Some students chose a cube to show which season they liked best, and put it into a graph. Look carefully at the graph and answer the questions in your jotter.

Our favourite season

Spring | Summer | Autumn | Winter

a) How many people prefer winter?

b) Which is the most popular season?

c) Which is the least popular season?

d) Read the statements below. Who is correct?

Six people said their favourite season is spring.

Six people said their favourite season is summer.

Nuria   Isla

e) Talk to a partner about what you can see in the graph.

Our favourite ice-cream

lemon

strawberry

Tell your partner two things about the graph.

14.3 Using data

We are learning to read information on charts and tables.

Before we start

a) Which colour ball is the **most** popular?

b) Which colour ball is the **least** popular?

I am learning to read data from a table, chart or diagram.

Let's learn

Graphs help us answer questions about information. The information is organised so that it is easier to read and compare.

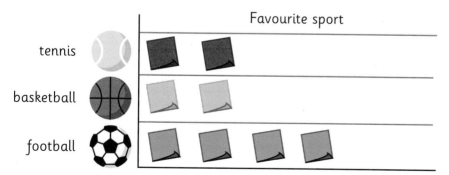

Favourite sport

tennis

basketball

football

Take time to make sure that objects are correctly placed on the graph so that it will show the correct information.

1) Some students from Class 1 have taken a cup of fruit juice to show which one they like best.

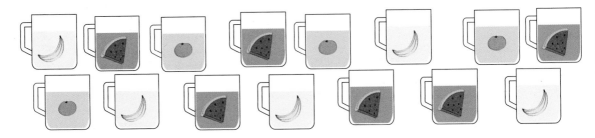

a) How many cups of banana juice are there?

b) How many students liked orange juice best?

c) Draw a cup with the flavour you would choose.

2) The students arranged their cups of juice to make a graph.

Our favourite juices

banana

watermelon

orange

a) What is the name of this fruit juice?

b) Talk to a partner about what you can see in the graph including the labels.

3) These students lined up by how old they are in years to make a graph.

a) Draw a picture of yourself in the correct line for your age.

b) What is the title?

c) What do the numbers at the side of the graph tell us?

Explain to someone or write it in your jotter.

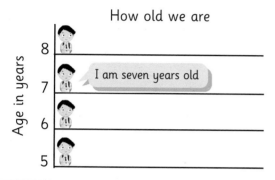

How old we are

Age in years

8

7 — I am seven years old

6

5

CHALLENGE! ....

Finlay is making a picture graph.

1) Does Finlay have more brothers or sisters?

My family, by Finlay   **Key:** ⋀ = 1 person

brothers	⋀	⋀					
sisters	⋀						

2) Draw a graph showing something about your family.

My family   **Key:** ⋀ = 1 person

Aunties							
Uncles							

15.1 Understanding chance

> We are learning to understand chance and uncertainty.

Before we start

Say if the following are *certain* or *uncertain*:

1) Grass is green.
2) The sky is blue.
3) It snows in winter.
4) Ice will melt.

> Things can be certain, possible or impossible.

Let's learn

Probability is the chance of something happening.

It can be certain. *Tuesday is before Wednesday.*

It can be possible. *It will rain tomorrow.*

It can be impossible. *I will get younger.*

Let's practise

1) Answer using *certain*, *possible* or *impossible*.
   Write your answers in your jotter.
   a) It will rain tomorrow.
   b) It will get dark tonight.
   c) I will make a mistake today.
   d) A cow will fly past the window.
   e) The winter will be warmer than the summer.
   f) I will learn something today.

2) Look at the marbles. Imagine we put them in a bag so you cannot see them. Is it certain, possible or impossible: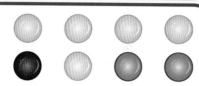
   a) You will pick a black marble out of the bag.
   b) You will pick a yellow marble out of the bag.
   c) You will pick a green marble out of the bag.
   d) You will pick a blue marble out of the bag.

3) Use your imagination.
   a) Write down two certain events.
   b) Write down two possible events.
   c) Write down two impossible events.

CHALLENGE! ....

In the story of Red Riding Hood, Red walks through the forest to see her Grandma.

Read the book to remember the story.

What is the chance that Red will see:

a) a tree?   b) a deer?   c) a wolf?   d) Santa Claus?

Now ask a partner to tell you something that is certain, possible or impossible to happen in the forest.